T0338634

CAMBRIDGE MONOGRAPHS ON
APPLIED AND COMPUTATIONAL
MATHEMATICS

Series Editors
M. ABLOWITZ, S. DAVIS, J. HINCH,
A. ISERLES, J. OCKENDON, P. OLVER

27 Difference Equations by Differential Equation Methods

The *Cambridge Monographs on Applied and Computational Mathematics* series reflects the crucial role of mathematical and computational techniques in contemporary science. The series publishes expositions on all aspects of applicable and numerical mathematics, with an emphasis on new developments in this fast-moving area of research.

State-of-the-art methods and algorithms as well as modern mathematical descriptions of physical and mechanical ideas are presented in a manner suited to graduate research students and professionals alike. Sound pedagogical presentation is a prerequisite. It is intended that books in the series will serve to inform a new generation of researchers.

A complete list of books in the series can be found at
www.cambridge.org/mathematics.
Recent titles include the following:

11. Generalized Riemann problems in computational fluid dynamics, *Matania Ben-Artzi & Joseph Falcovitz*

12. Radial basis functions, *Martin D. Buhmann*

13. Iterative Krylov methods for large linear systems, *Henk van der Vorst*

14. Simulating Hamiltonian dynamics, *Benedict Leimkuhler & Sebastian Reich*

15. Collocation methods for Volterra integral and related functional differential equations, *Hermann Brunner*

16. Topology for computing, *Afra J. Zomorodian*

17. Scattered data approximation, *Holger Wendland*

18. Modern computer arithmetic, *Richard Brent & Paul Zimmermann*

19. Matrix preconditioning techniques and applications, *Ke Chen*

20. Greedy approximation, *Vladimir Temlyakov*

21. Spectral methods for time-dependent problems, *Jan Hesthaven, Sigal Gottlieb & David Gottlieb*

22. The mathematical foundations of mixing, *Rob Sturman, Julio M. Ottino & Stephen Wiggins*

23. Curve and surface reconstruction, *Tamal K. Dey*

24. Learning theory, *Felipe Cucker & Ding Xuan Zhou*

25. Algebraic geometry and statistical learning theory, *Sumio Watanabe*

26. A practical guide to the invariant calculus, *Elizabeth Louise Mansfield*

Difference Equations by Differential Equation Methods

PETER E. HYDON
University of Surrey

CAMBRIDGE
UNIVERSITY PRESS

Shaftesbury Road, Cambridge CB2 8EA, United Kingdom

One Liberty Plaza, 20th Floor, New York, NY 10006, USA

477 Williamstown Road, Port Melbourne, VIC 3207, Australia

314–321, 3rd Floor, Plot 3, Splendor Forum, Jasola District Centre, New Delhi – 110025, India

103 Penang Road, #05–06/07, Visioncrest Commercial, Singapore 238467

Cambridge University Press is part of Cambridge University Press & Assessment,
a department of the University of Cambridge.

We share the University's mission to contribute to society through the pursuit of
education, learning and research at the highest international levels of excellence.

www.cambridge.org
Information on this title: www.cambridge.org/9780521878524

© Peter E. Hydon 2014

This publication is in copyright. Subject to statutory exception and to the provisions
of relevant collective licensing agreements, no reproduction of any part may take
place without the written permission of Cambridge University Press & Assessment.

First published 2014

A catalogue record for this publication is available from the British Library

ISBN   978-0-521-87852-4   Hardback

Cambridge University Press & Assessment has no responsibility for the persistence
or accuracy of URLs for external or third-party internet websites referred to in this
publication and does not guarantee that any content on such websites is, or will
remain, accurate or appropriate.

Every effort has been made in preparing this book to provide accurate and up-to-date
information which is in accord with accepted standards and practice at the time of
publication. Although case histories are drawn from actual cases, every effort has been
made to disguise the identities of the individuals involved. Nevertheless, the authors,
editors and publishers can make no warranties that the information contained herein
is totally free from error, not least because clinical standards are constantly changing
through research and regulation. The authors, editors and publishers therefore
disclaim all liability for direct or consequential damages resulting from the use of
material contained in this book. Readers are strongly advised to pay careful attention
to information provided by the manufacturer of any drugs or equipment that they plan
to use.

To

Alison,
Chris,
Rachel
and
Katy,

who didn't believe me
when I last said "Never again".

You were right.

Contents

vii

Preface

Difference equations are prevalent in mathematics, occurring in areas as disparate as number theory, control theory and integrable systems theory. They arise as mathematical models of discrete processes, as interesting dynamical systems, and as finite difference approximations to differential equations. Finite difference methods exploit the fact that differential calculus is a limit of the calculus of finite differences. It is natural to take this observation a step further and ask whether differential and difference equations share any common features. In particular, can they be solved by the same (or similar) methods?

Just over twenty years ago, a leading numerical analyst summarized the state of the art as follows: problems involving difference equations are an order of magnitude harder than their counterparts for differential equations. There were two major exceptions to this general rule. Linear ordinary difference equations behave similarly to their continuous counterparts. (Indeed, most of the best-known texts on difference equations deal mainly with linear and linearizable problems.) Discrete integrable systems are nonlinear, but have some underlying linear structures; they have much in common with continuous integrable systems, together with some interesting extra features.

Research since that time has transformed our understanding of more general difference equations and their solutions. The basic geometric structures that underpin difference equations are now known. From these, it has been possible to develop a plethora of systematic techniques for finding solutions, first integrals or conservation laws of a given difference equation. These look a little different to the corresponding methods for differential equations, mainly because the solutions of difference equations are not continuous. However, many techniques are widely applicable and most do not require the equation to have any special properties such as linearizability or integrability.

This book is intended to be an accessible introduction to techniques for general difference equations. The basic geometrical structures that lie behind these techniques are also described. My aim throughout has been to describe ideas that are applicable to all sufficiently well-behaved difference equations, without assuming prior knowledge of any special properties that an equation may possess. There is an exception: the last chapter includes material on Noether's Theorem and several related results, which apply only to difference equations that stem from a variational problem. As variational problems are commonplace, I think that many readers will find this material interesting and useful. I have not included material on the qualitative theory of difference equations. In particular, there is no discussion of stability or oscillation theory. I recommend Elaydi (2005) as an excellent introduction to these topics.

The book may be used as the basis for an advanced undergraduate or postgraduate course on difference equations; the main prerequisite is a working knowledge of solution methods for differential equations. It is also designed for self-study. Most of the material is presented informally, with few theorems and many worked examples; a triangle (▲) separates each example from the subsequent text. To help readers grasp the basic concepts and tools early on, each major idea is introduced in its simplest context, leaving generalizations until later. Every chapter concludes with a range of exercises of varying difficulty. Most are designed to enable readers to become proficient in using techniques; a few develop extensions of the core material. Where an exercise is particularly challenging, this is indicated by an asterisk (*).

Each chapter includes suggestions for further reading. These are works that will introduce readers to some topics in fields that are beyond the scope of this book (such as discrete integrable systems and geometric integration). I have not surveyed the literature on any specialized topic, however, in order to keep the book focused on methods that apply to most types of difference equations.

The first three chapters introduce the main ideas and methods in their simplest setting: ordinary difference equations (OΔEs). Together, these chapters include more than enough material for a single-semester course for advanced undergraduates. Chapter 1 surveys a range of methods for solving linear OΔEs. Its purpose is to give the beginner a rapid introduction to standard techniques, though some novel methods are also included. Chapter 2 introduces an indirect way to solve or simplify a given nonlinear OΔE, by first solving an associated linear problem. This approach, called symmetry analysis, lies at the heart of most exact methods for solving nonlinear ordinary differential equations. A dual approach is to construct first integrals; this too has a counterpart for OΔEs. There are various extensions of basic symmetry methods for OΔEs; some of the most useful ones are described in Chapter 3.

The pace is quickened somewhat in the second part of the book, which brings the reader to the leading edge of research. The focus is mainly on partial difference equations (PΔEs), starting in Chapter 4 with a detailed description of their underlying geometry. A key idea is that transforming a given PΔE (or writing it in different coordinates) may enable one to understand it better. In particular, one can identify circumstances under which an initial-value problem has a unique solution; this is a prerequisite for symmetry methods to work. Chapter 5 describes various methods for obtaining exact solutions of a given PΔE. These include some well-established techniques for linear PΔEs, together with symmetry analysis and other recent methods for nonlinear PΔEs.

Often, one is interested in properties that are shared by all solutions of a given partial differential equation, particularly conservation laws. Famously, Noether's Theorem uses symmetries of an underlying variational problem to generate conservation laws. There is also a way to construct conservation laws directly, whether or not a variational formulation is known. Chapter 6 describes PΔE analogues of Noether's Theorem and the direct method. Noether's paper on variational symmetries includes a second theorem, which identifies relationships between the Euler–Lagrange equations for gauge theories. Again, it turns out that there is a difference version of this result. The book concludes with a brief outline of the reason for the close analogy between conservation laws of differential and difference equations, with suggestions for further reading.

Difference equations have a well-deserved reputation for being tricky. Yet many of them are susceptible to differential equation techniques, provided that these are used with care. My aim has been to give the reader practical tools, coupled with sufficient geometrical understanding to use them correctly.

Peter E. Hydon

Acknowledgements

I thank my friends and colleages who have kindly read drafts of chapters and have suggested ways to improve them: Tim Grant, Rob Gray, Liz Mansfield, Gloria Mari-Beffa, Peter Olver, Linyu Peng, Jeevan Rai and Fiona Tomkinson. I particularly thank Wendy Atkins, David Tranah, Clare Dennison and my family for their encouragement and practical help.

1

Elementary methods for linear OΔEs

As Shakespeare says, if you're going to do a thing
you might as well pop right at it and get it over.

(P. G. Wodehouse, Very Good, Jeeves!)

This chapter is a rapid tour through the most useful methods for solving scalar linear ordinary difference equations (OΔEs). *En route*, we will see some differences between OΔEs and ordinary differential equations (ODEs). Yet what is most striking is how closely methods for linear OΔEs correspond to their counterparts for linear ODEs. Much of this similarity is due to linearity.

1.1 Basic definitions and notation

An OΔE or system of OΔEs has a single integer-valued independent variable, n, that can take any value within a domain $D \subset \mathbb{Z}$. A *scalar* OΔE (also called a *recurrence relation*) has just one dependent variable, u, which we shall assume is real-valued. The OΔE is *linear* if it can be written in the form

$$a_p(n)\,u(n+p) + a_{p-1}(n)\,u(n+p-1) + \cdots + a_0(n)\,u(n) = b(n), \qquad (1.1)$$

where p is a positive integer and each a_i is a given real-valued function. The *order* of the OΔE at n is the difference between the highest and lowest arguments of u in (1.1). A point $n \in D$ is a *regular point* of the OΔE if $a_p(n)$ and $a_0(n)$ are both nonzero; otherwise it is *singular*. The OΔE (1.1) is of order p only at regular points; it is of lower order at singular points.

Example 1.1 The following OΔE has two singular points:

$$(n - 1)^2 u(n + 2) + 2\,u(n + 1) + (1 - n^2)\,u(n) = 0, \qquad n \in \mathbb{Z}.$$

At $n = -1$, the OΔE reduces to the first-order equation $4\,u(1) + 2\,u(0) = 0$. At $n = 1$, it amounts to $2\,u(2) = 0$, which is not even a difference equation. ▲

Example 1.2   As an extreme example, consider the OΔE

$$(1 + (-1)^n)\, u(n + 2) - 2u(n + 1) + (1 - (-1)^n)\, u(n) = 0, \qquad n \in \mathbb{Z}. \quad (1.2)$$

Although this looks like a second-order OΔE, *every* point is singular! Furthermore, (1.2) yields exactly the same equation when $n = 2m$ as it does when $n = 2m + 1$, namely

$$u(2m+2) = u(2m+1). \qquad \blacktriangle$$

Given an OΔE (1.1), we call D a *regular domain* if it is a set of consecutive regular points. Suppose that $\mathbb{Z}$ is a regular domain, so that $a_p(n)$ and $a_0(n)$ are nonzero for every integer n. Suppose also that, for some n_0, the p consecutive values $u(n_0), \dots, u(n_0 + p - 1)$ are known. Then one can calculate $u(n_0 + p)$ by setting $n = n_0$ in (1.1). Repeating this process, using $n = n_0 + 1, n_0 + 2, \dots$ in turn, produces $u(n_0 + i)$ for all $i > p$. The remaining values of u can also be obtained by setting $n = n_0 - 1, n_0 - 2$, and so on. Thus p (arbitrary) consecutive conditions determine a unique solution of the OΔE. This result holds for any regular domain; consequently, the general solution of every p^{th}-order linear OΔE on a regular domain depends upon p arbitrary constants.

When the domain is not regular, various oddities may occur. For instance, the general solution of (1.2) depends on infinitely many arbitrary constants, because for each $m \in \mathbb{Z}$, one of the pair $\{u(2m+1), u(2m+2)\}$ must be given in order to determine the other. To avoid having to deal separately with singularities, we will ensure that D is regular from here on, setting $a_p(n) = 1$ without loss of generality; under these conditions, we describe the OΔE (1.1) as being in *standard form*.

For a pth-order linear ODE,

$$y^{(p)}(x) + a_{p-1}(x)\, y^{(p-1)}(x) + \cdots + a_0(x)\, y(x) = b(x),$$

the usual convention suppresses the argument x where this can be assumed, leading to a slightly shorter expression:

$$y^{(p)} + a_{p-1}(x)\, y^{(p-1)} + \cdots + a_0(x)\, y = b(x).$$

Similarly, it is helpful to write the OΔE (1.1), in standard form, as

$$u_p + a_{p-1}(n)\, u_{p-1} + \cdots + a_0(n)\, u = b(n), \qquad (1.3)$$

where u and u_i are shorthand for $u(n)$ and $u(n+i)$, respectively. Suppressing the independent variable(s) saves considerable space, particularly for partial difference equations[1], so we will do this for all *unknown* functions of n. If a

---

[1]   For instance, if u depends on $\mathbf{n} = (n^1, n^2, n^3, n^4)$ then $u(n^1 + 1, n^2 + 3, n^3 + 2, n^4 + 1)$ is written as $u_{1,3,2,1}$.

function f is given, or is assumed to be known, we use $f(n)$, $f(n + 1)$, and so on. In particular, a solution of the OΔE (1.3) will be an expression of the form

$$u = f(n),$$

where, for each $n \in D$, $u(n) = f(n)$ satisfies (1.1) with $a_p(n) = 1$.

If $b(n) = 0$ for each $n \in D$ then (1.3) is *homogeneous*; otherwise, (1.3) is *inhomogeneous* and its associated homogeneous equation is

$$u_p + a_{p-1}(n) u_{p-1} + \cdots + a_0(n) u = 0. \tag{1.4}$$

These definitions correspond to those that are used for linear ODEs. Just as for ODEs, there is a principle of linear superposition: if $u = f_1(n)$ and $u = f_2(n)$ are any two solutions of (1.4) then

$$u = c_1 f_1(n) + c_2 f_2(n) \tag{1.5}$$

is a solution for all constants c_1 and c_2. Henceforth, the notation c_i and $\tilde{c}_i$ will be reserved for arbitrary constants. (Sometimes it is convenient to replace one set of arbitrary constants, $\{c_i\}$, by another set, $\{\tilde{c}_i\}$.)

Suppose that p solutions of (1.4), $u = f_i(n)$, $i = 1, \ldots, p$, are linearly independent[2] on D, which means that

$$\sum_{i=1}^{p} c_i f_i(n) = 0, \text{ for all } n \in D \text{ if and only if } c_1 = c_2 = \cdots = c_p = 0.$$

Then the principle of linear superposition implies that

$$u = \sum_{i=1}^{p} c_i f_i(n) \tag{1.6}$$

is a solution of (1.4) for each choice of values for the p arbitrary constants c_i; consequently (1.6) is the general solution of (1.4). If, in addition, $u = g(n)$ is any particular solution of a given inhomogeneous OΔE (1.3), the general solution of that OΔE is

$$u = g(n) + \sum_{i=1}^{p} c_i f_i(n). \tag{1.7}$$

These results are proved in the same way as their counterparts for ODEs.

---

[2] A simple test for linear independence is given in Exercise 1.1.

1.2 The simplest OΔEs: solution by summation

The starting-point for the solution of ODEs is the Fundamental Theorem of Calculus, namely

$$\int_{x=x_0}^{x_1} y'(x)\, \mathrm{d}x = y(x_1) - y(x_0).$$

A similar result holds for difference equations. Define the action of the *forward difference operator* Δ_n on any function $f(n)$ by

$$\Delta_n f(n) \equiv f(n+1) - f(n), \qquad \text{for all } n \in \mathbb{Z}. \tag{1.8}$$

In particular,

$$\Delta_n u \equiv u_1 - u \quad \text{and} \quad \Delta_n u_i \equiv u_{i+1} - u_i, \quad i \in \mathbb{Z}. \tag{1.9}$$

By summing consecutive differences, we obtain

$$\sum_{k=n_0}^{n_1-1} \Delta_k f(k) = \sum_{k=n_0}^{n_1-1} (f(k+1) - f(k)) = f(n_1) - f(n_0). \tag{1.10}$$

This very useful result is known as the *Fundamental Theorem of Difference Calculus*. We can use it immediately to solve OΔEs of the form

$$u_1 - u = b(n), \qquad n \geq n_0. \tag{1.11}$$

Replacing n by k in (1.11) and (f, n_1) by (u, n) in (1.10) yields

$$u = u(n_0) + \sum_{k=n_0}^{n-1} b(k); \tag{1.12}$$

we adopt the convention that a sum is zero if its lower limit exceeds its upper limit, which occurs here only when $n = n_0$. The OΔE (1.11) is recovered by applying the forward difference operator Δ_n to (1.12):

$$\Delta_n u = \left(u(n_0) + \sum_{k=n_0}^{n} b(k) \right) - \left(u(n_0) + \sum_{k=n_0}^{n-1} b(k) \right) = b(n).$$

If the initial condition $u(n_0)$ is known, (1.12) is the unique solution of (1.11) that satisfies this initial condition; otherwise $u(n_0)$ in (1.12) can be replaced by an arbitrary constant c_1, which yields the general solution of (1.11). For instance, the general solution of the simplest OΔE, $\Delta_n u = 0$, on any regular domain is $u = c_1$.

1.2.1 Summation methods

The solution (1.12) involves an unevaluated sum over multiple points k. If the sum is written as a function that is evaluated at n only, the solution is said to be in *closed form*. To obtain a closed-form solution of the OΔE (1.11), we need to find an *antidifference* (or indefinite sum) of the function $b(k)$, which is any function $B(k)$ that satisfies

$$\Delta_k B(k) = b(k) \quad \text{for every } k \in D. \tag{1.13}$$

If $B(k)$ is an antidifference of $b(k)$, so is $B(k) + c$ for any real constant c. Moreover, $u = B(n)$ is a particular solution of (1.11), so the general solution of this OΔE is

$$u = B(n) + \tilde{c}_1.$$

Indeed, if we substitute (1.13) into the sum in (1.12) and use the Fundamental Theorem of Difference Calculus, we obtain

$$u = B(n) - B(n_0) + u(n_0).$$

Summation is the difference analogue of integration, so antidifferences are as useful for solving OΔEs as antiderivatives (or indefinite integrals) are for solving ODEs. Table 1.1 lists some elementary functions and their antidifferences; these are sufficient to deal with the most commonly occurring sums. The functions $k^{(r)}$ in Table 1.1 are defined as follows:

$$k^{(r)} = \begin{cases} \dfrac{k!}{(k-r)!}, & k \geq r, \\ 0, & k < r. \end{cases} \tag{1.14}$$

The formula for the antidifference of $k^{(r)}$ looks very like the formula for the indefinite integral of x^r; it is easy to evaluate sums of these functions, as we are on familiar ground. In particular, we can sum all polynomials in (non-negative) k, because these can be decomposed into a sum of the functions

$$k^{(r)} = k(k-1)\cdots(k-r+1), \qquad 0 \leq r \leq k.$$

Example 1.3   Evaluate $\displaystyle\sum_{k=1}^{n-1} k^2$ and $\displaystyle\sum_{k=1}^{n-1} k^3$ in closed form.

Solution: Write $k^2 = k(k-1) + k = k^{(2)} + k^{(1)}$; then row **3** of Table 1.1 yields

$$k^2 = \Delta_k \left(\frac{1}{3} k^{(3)} + \frac{1}{2} k^{(2)} \right).$$

Table 1.1 *Antidifferences of some elementary functions:* $\Delta_k(B(k)) = b(k)$.

	Function $b(k)$	Antidifference $B(k)$
1	1	k
2	k	$k(k-1)/2$
3	$k^{(r)}, \ (r \neq -1, k \geq 0)$	$k^{(r+1)}/(r+1)$
4	$a^k, \ (a \neq 1)$	$a^k/(a-1)$
5	$\cos(ak+b), \ (a \not\equiv 0, \bmod 2\pi)$	$\sin(ak+b-a/2)/(2\sin(a/2))$
6	$\sin(ak+b), \ (a \not\equiv 0, \bmod 2\pi)$	$-\cos(ak+b-a/2)/(2\sin(a/2))$

Now sum to get

$$\sum_{k=1}^{n-1} k^2 = \left[\frac{1}{3}k^{(3)} + \frac{1}{2}k^{(2)} \right]_{k=1}^{n} = \frac{1}{6}n(n-1)(2n-1).$$

Similarly, the decomposition $k^3 = k^{(3)} + 3k^{(2)} + k^{(1)}$ yields

$$\sum_{k=1}^{n-1} k^3 = \left[\frac{1}{4}k^{(4)} + k^{(3)} + \frac{1}{2}k^{(2)} \right]_{k=1}^{n} = \frac{1}{4}n^2(n-1)^2. \qquad \blacktriangle$$

The functions $k^{(r)}$ are also useful for summing some rational polynomials because, for $r < 0$,

$$k^{(r)} = \frac{1}{(k+1)(k+2)\cdots(k-r)}. \qquad (1.15)$$

Example 1.4   Calculate $\displaystyle\sum_{k=1}^{n-1} \frac{k}{(k+1)(k+2)(k+3)}$.

Solution: First decompose the summand as follows:

$$\frac{k}{(k+1)(k+2)(k+3)} = \frac{1}{(k+1)(k+2)} - \frac{3}{(k+1)(k+2)(k+3)} = k^{(-2)} - 3k^{(-3)}.$$

Then use row **3** of Table 1.1 to obtain

$$\sum_{k=1}^{n-1} \frac{k}{(k+1)(k+2)(k+3)} = \left[-k^{(-1)} + \frac{3}{2}k^{(-2)} \right]_{k=1}^{n} = \frac{n(n-1)}{4(n+1)(n+2)}. \qquad \blacktriangle$$

The method used in this example works for any rational polynomial of the form $k^{(r)}P_m(k)$, where $r \leq -2$ and $P_m(k)$ is a polynomial of degree $m \leq -r-2$. The reason why $P_m(k)$ must be at least two orders lower than $1/k^{(r)}$ is that terms proportional to $k^{(-1)}$ cannot be evaluated in closed form. The sums

$$H_n = \sum_{k=0}^{n-1} k^{(-1)} = \sum_{k=1}^{n} \frac{1}{k}$$

are called *harmonic numbers*. For large n, the harmonic numbers are approximated to $O(n^{-2})$ by $H_n \approx \ln(n) + \gamma + 1/(2n)$; here $\gamma \approx 0.5772$ is the Euler constant. A better estimate, whose error is $O(n^{-4})$, is

$$H_n \approx \frac{\ln(n+1) + \ln(n)}{2} + \gamma + \frac{1}{6n(n+1)}.$$

A sum whose summand is a periodic function of k may be expressed in closed form as follows. Calculate the contribution per period and multiply this by the number of complete periods in the range; then add any remaining terms. If the period is small, this is often the simplest way of evaluating such sums. Alternatively, one can write the periodic function as a sum of sines and cosines and then use rows **5** and **6** of Table 1.1.

Example 1.5 Calculate $\displaystyle\sum_{k=n}^{0} (-1)^{k(k-1)/2}$, where $n < 0$.

Solution: Note that $(-1)^{k(k-1)/2} = \sqrt{2}\sin\left((2k+1)\pi/4\right)$; so row **6** gives

$$\sum_{k=n}^{0} (-1)^{k(k-1)/2} = \left[-\cos\left(k\pi/2\right)\right]_{k=n}^{1} = \cos\left(n\pi/2\right). \qquad \blacktriangle$$

These methods of summation can be combined to deal with summands that are products of powers, polynomials and periodic functions. It would be convenient if there were an analogue of the Leibniz product rule (which leads directly to the formula for integration by parts). Unlike the differential operator d/dx, however, the forward difference operator does not satisfy the Leibniz product rule, because

$$\Delta_n\{f(n)\,g(n)\} = f(n+1)\,g(n+1) - f(n)\,g(n)$$
$$\neq \Delta_n\{f(n)\}g(n) + f(n)\,\Delta_n\{g(n)\}.$$

Instead, the following modified Leibniz rule holds:

$$\Delta_n\{f(n)\,g(n)\} = \Delta_n\{f(n)\}\,g(n+1) + f(n)\,\Delta_n g(n). \qquad (1.16)$$

This leads to the extraordinarily useful *summation by parts* formula,

$$\sum_{k=n_0}^{n_1-1} f(k)\,\Delta_k g(k) = \left[f(k)\,g(k)\right]_{k=n_0}^{n_1} - \sum_{k=n_0}^{n_1-1} \{\Delta_k f(k)\}\,g(k+1). \qquad (1.17)$$

Example 1.6   Calculate $\sum_{k=1}^{n-1} ka^k$, where $a \neq 1$.

Solution: Substitute $f(k) = k$, $g(k) = a^k/(a-1)$ into (1.17) to obtain

$$\sum_{k=1}^{n-1} ka^k = \left[\frac{ka^k}{a-1}\right]_{k=1}^{n} - \sum_{k=1}^{n-1}\frac{a^{k+1}}{a-1} = \left[\frac{ka^k}{a-1} - \frac{a^{k+1}}{(a-1)^2}\right]_{k=1}^{n} = \frac{na^n}{a-1} - \frac{a^{n+1}-a}{(a-1)^2}.$$

▲

These elementary summation techniques are sufficient to deal with most simple problems; however, there are many functions whose sum cannot be expressed in closed form. Nevertheless, we will regard an OΔE as being solved when its general solution is given, even if this is in terms of one or more unevaluated sums.

1.2.2 The summation operator

So far, we have restricted attention to OΔEs for which $n \geq n_0$. If solutions are sought for all $n \in \mathbb{Z}$, or for $n < n_0$, it is also necessary to solve

$$u_1 - u = b(n), \qquad n < n_0. \tag{1.18}$$

To do this, use (1.10) with n_0 and n_1 replaced by n and n_0 respectively, which yields

$$u = u(n_0) - \sum_{k=n}^{n_0-1} b(k). \tag{1.19}$$

It is helpful to combine (1.12) and (1.19) by defining the *summation operator* σ_k as follows:

$$\sigma_k\{f(k); n_0, n_1\} = \begin{cases} \displaystyle\sum_{k=n_0}^{n_1-1} f(k), & n_1 > n_0; \\[2mm] 0, & n_1 = n_0; \\[2mm] \displaystyle -\sum_{k=n_1}^{n_0-1} f(k), & n_1 < n_0. \end{cases} \tag{1.20}$$

This operator satisfies the identity

$$\sigma_k\{f(k); n_0, n+1\} = \sigma_k\{f(k); n_0, n\} + f(n). \tag{1.21}$$

So, given $u(n_0)$, the solution of

$$\Delta_n u = b(n), \qquad n \in \mathbb{Z}, \tag{1.22}$$

that satisfies the initial condition is

$$u = u(n_0) + \sigma_k\{b(k); n_0, n\}. \tag{1.23}$$

If no initial condition is prescribed, the general solution of (1.22) is

$$u = c_1 + \sigma_k\{b(k); n_0, n\}, \tag{1.24}$$

where n_0 is any convenient integer in the domain. In accordance with the principle of linear superposition, $u = c_1$ is the general solution of the associated homogeneous equation, $\Delta_n u = 0$, and $u = \sigma_k\{b(k); n_0, n\}$ is a particular solution of (1.22).

1.3 First-order linear OΔEs

The standard form of any first-order linear homogeneous OΔE is

$$u_1 + a(n)\,u = 0, \qquad a(n) \neq 0, \quad n \in D. \tag{1.25}$$

As in the last section, we begin by solving the OΔE on $D = \{n \in \mathbb{Z} : n \geq n_0\}$ before generalizing the result to arbitrary regular domains. To find u for $n > n_0$, replace n by $n - 1$ in (1.25) and rearrange the OΔE as follows:

$$u = -a(n - 1)\,u_{-1}. \tag{1.26}$$

Then replace n by $n - 1, n - 2, \ldots, n_0 + 1$ successively in (1.26) to obtain

$$
\begin{aligned}
u &= (-1)^2 a(n - 1)\,a(n - 2)\,u_{-2} \\
&= (-1)^3 a(n - 1)\,a(n - 2)\,a(n - 3)\,u_{-3} \\
&\;\;\vdots \\
&= (-1)^{n-n_0} a(n - 1)\,a(n - 2)\,a(n - 3)\cdots a(n_0)\,u(n_0).
\end{aligned}
$$

Therefore the solution of (1.25) for $n > n_0$ is

$$u = u(n_0) \prod_{k=n_0}^{n-1} (-a(k)) = u(n_0)(-1)^{n-n_0} \prod_{k=n_0}^{n-1} a(k). \tag{1.27}$$

If no initial condition is prescribed, $u(n_0)$ is replaced by an arbitrary constant.

Example 1.7 Solve the initial-value problem

$$u_1 - 3u = 0, \qquad n \geq 0, \qquad \text{subject to} \quad u(0) = 2.$$

Solution: Substitute $a(n) = -3$, $n_0 = 0$ and $u(n_0) = 2$ into (1.27) to obtain

$$u = 2 \prod_{k=0}^{n-1} 3 = 2 \times 3^n. \qquad \blacktriangle$$

Example 1.8　Let α be a positive constant. Find the general solution of

$$u_1 - (n + \alpha)u = 0, \qquad n \geq 0.$$

Solution: Here $a(n) = -(n + \alpha)$ and $n_0 = 0$, so (1.27) yields

$$u = c_1 \prod_{k=0}^{n-1} (k + \alpha) = \frac{c_1 \Gamma(n + \alpha)}{\Gamma(\alpha)}.$$

Here $\Gamma(z)$ is the *gamma function* (the analytic continuation of the factorial function to the complex plane with the non-positive integers $0, -1, -2, \ldots$ deleted). The gamma function satisfies

$$\Gamma(z + 1) = z\Gamma(z); \qquad (1.28)$$

for $\mathrm{Re}\{z\} > 0$, it is defined by the integral

$$\Gamma(z) = \int_0^\infty t^{z-1} e^{-t}\, dt. \qquad (1.29)$$

A simple integration by parts shows that the definition (1.29) is consistent with (1.28). The gamma function may also be evaluated in the region $\mathrm{Re}\{z\} \leq 0$, except at the deleted points. This is done by using (1.28) repeatedly to find $\Gamma(z)$ in terms of $\Gamma(z + N)$, where $N \in \mathbb{N}$ is large enough to make $\mathrm{Re}\{z + N\}$ positive. Two immediate consequences of (1.28) and (1.29) are the useful identities

$$\Gamma(n + 1) = n! \quad \text{for all } n \in \mathbb{N}, \qquad\qquad \Gamma(1/2) = \sqrt{\pi}. \qquad ▲$$

It is worth observing that (1.27) is closely related to the formula (1.12) for inverting the forward difference operator. For simplicity, suppose that $a(n) < 0$ for every $n \geq n_0$ and that $u(n_0) > 0$; this guarantees that u is positive throughout the domain. Then the logarithm of (1.26), with n replaced by $n+1$, amounts to the OΔE

$$\Delta_n(\ln u) = \ln(-a(n)), \qquad n \geq n_0. \qquad (1.30)$$

From (1.12), the general solution of (1.30) is

$$\ln u = \ln(u(n_0)) + \sum_{k=n_0}^{n-1} \ln(-a(k)) = \ln\left(u(n_0) \prod_{l=n_0}^{n-1} (-a(l))\right),$$

which is the logarithm of (1.27). In other words, we could have solved (1.25) by transforming it to an OΔE of a type that we already know how to solve. The same approach lies at the heart of many methods for OΔEs and ODEs alike: transform them to something simpler. Later in this chapter, we will consider transformations that are particularly useful for linear OΔEs. Suitable transformations of nonlinear OΔEs may be found systematically, using a technique that is described in the next chapter.

The summation operator σ_k was introduced to allow a general treatment of all regular domains. For the same reason, it is convenient to introduce a *product operator* π_l, which is defined by

$$\pi_l\{f(l); n_0, n_1\} = \begin{cases} \displaystyle\prod_{l=n_0}^{n_1-1} f(l), & n_1 > n_0; \\ 1, & n_1 = n_0; \\ \displaystyle\prod_{l=n_1}^{n_0-1} 1/f(l), & n_1 < n_0. \end{cases} \tag{1.31}$$

This is similar in form to the summation operator. It satisfies the identity

$$\pi_l\{f(l); n_0, n+1\} = f(n)\,\pi_l\{f(l); n_0, n\}. \tag{1.32}$$

The general solution of the homogeneous OΔE (1.25) on an arbitrary regular domain D is

$$u = c_1\pi_l\{-a(l); n_0, n\}. \tag{1.33}$$

As before, n_0 can be chosen to be any convenient integer in D.

Inhomogeneous first-order linear OΔEs, like their ODE counterparts, can be solved by the *integrating factor* method. Simply divide the OΔE

$$u_1 + a(n)\,u = b(n), \qquad a(n) \neq 0, \tag{1.34}$$

by $\pi_l\{-a(l); n_0, n+1\}$ and use the identity (1.32) to obtain

$$\Delta_n\left(\frac{u}{\pi_l\{-a(l); n_0, n\}}\right) = \frac{u_1 + a(n)\,u}{\pi_l\{-a(l); n_0, n+1\}} = \frac{b(n)}{\pi_l\{-a(l); n_0, n+1\}}.$$

Now apply the summation operator to obtain the general solution of (1.34):

$$u = \pi_l\{-a(l); n_0, n\}\left(u(n_0) + \sigma_k\left\{\frac{b(k)}{\pi_l(-a(l); n_0, k+1)}; n_0, n\right\}\right). \tag{1.35}$$

Although this result looks somewhat daunting at first sight, it is quite easy to use in practice. The advantage of using σ_k and π_l is that the result can be stated in full generality, without the need to consider the details of the domain D.

Example 1.9 Solve the initial-value problem

$$u_1 - 3u = 3^n n, \qquad n \geq 0, \qquad \text{subject to } u(0) = 2.$$

Solution: Here $a(n) = -3$, $n_0 = 0$ and $n \geq n_0$, so we obtain (as in Example 1.7)

$$\pi_l\{-a(l); n_0, n\} = \pi_l\{3; 0, n\} = \prod_{l=0}^{n-1} 3 = 3^n.$$

Substituting this, $b(n) = 3^n n$ and $u(0) = 2$ into (1.35) yields the solution

$$u = 3^n (2 + \sigma_k \{k/3; 0, n\}) = 3^n \left(2 + \sum_{k=0}^{n-1} k/3\right) = 3^n (2 + n(n-1)/6). \quad \blacktriangle$$

Example 1.10   Find the general solution of

$$u_1 - (-1)^n u = (-1)^n, \qquad n \leq 1.$$

Solution: Here $a(n) = -(-1)^n$, $n_0 = 1$ and $n \leq n_0$, so

$$\pi_l \{-a(l); n_0, n\} = \prod_{l=n}^{0} \frac{1}{(-1)^l} = (-1)^{n(n-1)/2}.$$

Therefore (1.35) yields the general solution

$$u = (-1)^{n(n-1)/2} \left(c_1 + \sigma_k \left\{\frac{(-1)^k}{(-1)^{k(k+1)/2}}; 1, n\right\}\right)$$

$$= (-1)^{n(n-1)/2} \left(c_1 - \sum_{k=n}^{0} (-1)^{k(k-1)/2}\right)$$

$$= (-1)^{n(n-1)/2} (c_1 - \cos(n\pi/2)). \qquad \blacktriangle$$

1.4 Reduction of order

The integrating factor method cannot be used for OΔEs of order $p \geq 2$. However, it has a generalization called *reduction of order*, which enables one to solve a pth-order linear OΔE by solving a linear OΔE of order $p - 1$. This method has a counterpart for ODEs; it is useful to review the ODE method before we consider OΔEs. Given a linear ODE of order p,

$$y^{(p)} + a_{p-1}(x) y^{(p-1)} + \cdots + a_0(x) y = \beta(x),$$

find a nonzero solution, $y = f(x)$, of the associated homogeneous equation. Then substitute $y = f(x) z$ into the original ODE; this produces a 'reduced' linear ODE for $z' = dz/dx$ that is of order $p - 1$. The reduced ODE is homogeneous if and only if the original ODE is homogeneous. Suppose that the reduced ODE can be solved (possibly by further reductions of order); its general solution will be of the form

$$z' = g(x; c_1, \ldots, c_{p-1}),$$

which can be integrated to find z. Then the general solution of the original ODE is

$$y = f(x) \left[c_p + \int^x g(x; c_1, \ldots, c_{p-1}) \, dx \right].$$

With minor modifications, the same method can be applied to linear OΔEs. In §1.3, we used the integrating factor method to rewrite

$$u_1 + a(n) \, u = b(n), \qquad a(n) \neq 0, \tag{1.36}$$

as

$$\Delta_n \left(\frac{u}{\pi_l\{-a(l); n_0, n\}} \right) = \frac{b(n)}{\pi_l\{-a(l); n_0, n + 1\}},$$

which can be summed to yield u. Equivalently, we could have introduced a new variable v that satisfies

$$u = \pi_l\{-a(l); n_0, n\} \, v,$$

to obtain an OΔE that is immediately summable:

$$\Delta_n v = \frac{b(n)}{\pi_l\{-a(l); n_0, n + 1\}}. \tag{1.37}$$

The key to this method is that we know a nonzero solution of the associated homogeneous equation,

$$u_1 + a(n) \, u = 0,$$

namely $u = \pi_l\{-a(l); n_0, n\}$. Therefore the substitution that transforms (1.36) to the simpler OΔE (1.37) is of the same type as is used in reduction of order for ODEs.

Now consider the effect of this approach on a second-order linear OΔE,

$$u_2 + a_1(n) \, u_1 + a_0(n) \, u = b(n), \qquad a_0(n) \neq 0. \tag{1.38}$$

Let $u = f(n)$ be any solution of the associated homogeneous OΔE,

$$u_2 + a_1(n) \, u_1 + a_0(n) \, u = 0, \tag{1.39}$$

such that $f(n)$ and $f(n + 2)$ are nonzero for every $n \in D$. Substitute $u = f(n) \, v$ into (1.38) to obtain the following linear OΔE for v:

$$f(n + 2) \, v_2 + a_1(n) \, f(n + 1) \, v_1 + a_0(n) \, f(n) \, v = b(n). \tag{1.40}$$

At first sight, (1.40) does not look any simpler than the original OΔE. However, the substitution

$$w = \Delta_n v = v_1 - v \tag{1.41}$$

reduces (1.40) to a first-order linear OΔE:

$$f(n + 2) w_1 - a_0(n) f(n) w = b(n). \tag{1.42}$$

The requirement that $f(n)$ and $f(n + 2)$ are nonzero on D ensures that D is a regular domain for (1.42). Now use the integrating factor method to solve the reduced OΔE (1.42); if the solution is

$$w = g(n; c_1),$$

all that remains is to use summation to solve (1.41) for v. Thus the solution of the original OΔE (1.38) is

$$u = f(n)(c_2 + \sigma_k \{g(k; c_1); n_0, n\}). \tag{1.43}$$

Example 1.11   Solve the OΔE

$$u_2 - (n + 2) u_1 + (n + 2) u = 0, \qquad n \geq 1, \tag{1.44}$$

using the fact that it has a solution $u = n$.

Solution: Substitute $u = nv$ into (1.44) and divide by $n + 2$ to obtain

$$v_2 - (n + 1)v_1 + nv = 0.$$

The substitution (1.41) reduces this to

$$w_1 - nw = 0,$$

whose general solution is

$$w = c_1 \pi_l \{l; 1, n\} = c_1(n - 1)!.$$

Therefore (1.43) yields, with $f(n) = n$,

$$u = n(c_2 + c_1 \sigma_k \{(k - 1)! ; 1, n\}) = c_1 \left(n \sum_{k=1}^{n-1} (k - 1)! \right) + c_2 n. \qquad \blacktriangle$$

Example 1.12   Find the solution of

$$u_2 + (1/n) u_1 - (1 + 1/n) u = 4(n + 1), \qquad n \geq 1, \tag{1.45}$$

that satisfies the initial conditions $u(1) = 1$, $u(2) = -1$.

Solution: The first task is to try to find a nonzero solution of the associated homogeneous OΔE

$$u_2 + (1/n) u_1 - (1 + 1/n) u = 0.$$

One solution is obvious, namely $u = 1$. Therefore the recipe for reduction of order gives $v = u$, and so (1.45) can be reduced immediately by the substitution $w = u_1 - u$. The reduced OΔE is

$$w_1 + (1 + 1/n) w = 4(n + 1). \tag{1.46}$$

Now apply the integrating factor method, which leads to the result

$$w = c_1 n (-1)^n + 2n.$$

(This is left as an exercise.) Finally, apply the summation operator to obtain the general solution of (1.45):

$$u = c_2 + c_1 \sigma_k \left\{ k(-1)^k; 1, n \right\} + \sigma_k \{2k; 1, n\}$$
$$= \tilde{c}_2 + \tilde{c}_1 (2n - 1)(-1)^{n+1} + n(n - 1).$$

The initial conditions yield a pair of simultaneous equations for $\tilde{c}_1$ and $\tilde{c}_2$:

$$n = 1: \qquad \tilde{c}_1 + \tilde{c}_2 = 1,$$
$$n = 2: \qquad -3\tilde{c}_1 + \tilde{c}_2 + 2 = -1.$$

Hence $\tilde{c}_1 = 1$, $\tilde{c}_2 = 0$, and so the solution of the initial-value problem is

$$u = (2n - 1)(-1)^{n+1} + n(n - 1). \qquad \blacktriangle$$

Although we have considered only second-order linear OΔEs, the method of reduction of order is equally applicable to linear OΔEs of order $p \geq 3$. If $u = f(n)$ is a solution of the associated homogeneous equation such that $f(n)$ and $f(n + p)$ are nonzero on D, the substitutions

$$u = f(n)v, \qquad w = v_1 - v, \tag{1.47}$$

reduce the original OΔE to a linear OΔE of order $p - 1$ for w. If the reduced OΔE can be solved, u may be found exactly as for second-order OΔEs.

Reduction of order enables us to solve an inhomogeneous linear OΔE provided that we can find sufficiently many solutions of the associated homogeneous equation. Therefore the remainder of this chapter examines various methods that yield solutions of linear homogeneous OΔEs. For simplicity, we will restrict attention to second-order OΔEs. Nevertheless, the methods are equally applicable to higher-order problems.

1.5 OΔEs with constant coefficients

The simplest class of linear homogeneous OΔEs are those for which all of the coefficients $a_i(n)$ are constants. In particular, any first-order OΔE of the form

$$u_1 - \lambda u = 0 \tag{1.48}$$

(where λ is a nonzero real constant) has the general solution

$$u = c_1 \lambda^n. \tag{1.49}$$

This result enables us to solve higher-order OΔEs with constant coefficients. Consider the second-order OΔE

$$u_2 + \alpha u_1 + \beta u = 0, \qquad \beta \neq 0, \tag{1.50}$$

where α and β are real constants. This OΔE has solutions of the form (1.49) for arbitrary c_1 precisely when λ is a root of the *auxiliary equation*:

$$\lambda^2 + \alpha \lambda + \beta = 0. \tag{1.51}$$

Hence

$$\lambda = \tfrac{1}{2}\left(-\alpha \pm \sqrt{\alpha^2 - 4\beta}\right).$$

If we seek real solutions of (1.50), there are three cases to consider.

Case I: Distinct real roots. If $\alpha^2 - 4\beta > 0$, the auxiliary equation has two distinct real roots, namely

$$\lambda_1 = \tfrac{1}{2}\left(-\alpha + \sqrt{\alpha^2 - 4\beta}\right), \qquad \lambda_2 = \tfrac{1}{2}\left(-\alpha - \sqrt{\alpha^2 - 4\beta}\right).$$

Therefore (1.49) is a solution with either λ_1 or λ_2 replacing λ. By the principle of linear superposition,

$$u = c_1 \lambda_1^n + c_2 \lambda_2^n \tag{1.52}$$

is a real solution of (1.50) for each $c_1, c_2 \in \mathbb{R}$. Moreover, this is the general solution, as it is an arbitrary linear combination of two linearly independent solutions.

Example 1.13   Find the general solution of

$$u_2 - 4u_1 + 3u = 0, \qquad n \in \mathbb{Z}.$$

Solution: The roots of the auxiliary equation,

$$\lambda^2 - 4\lambda + 3 = 0,$$

are $\lambda_1 = 3$ and $\lambda_2 = 1$. Therefore the general solution of the OΔE is

$$u = c_1 3^n + c_2. \qquad\qquad\qquad \blacktriangle$$

Case II: Distinct complex roots. If $\alpha^2 - 4\beta < 0$, the roots of the auxiliary equation are complex conjugates:

$$\lambda_1 = \tfrac{1}{2}\left(-\alpha + i\sqrt{4\beta - \alpha^2}\right), \qquad \lambda_2 = \tfrac{1}{2}\left(-\alpha - i\sqrt{4\beta - \alpha^2}\right).$$

As the roots are distinct, the general solution could be written in the form (1.52); this is real if and only if c_1 and c_2 are complex conjugate constants. It is usually more convenient to use real constants, which can be achieved by writing the roots in polar form:

$$\lambda_1 = re^{i\theta}, \qquad \lambda_2 = re^{-i\theta}, \tag{1.53}$$

where

$$r = \sqrt{\beta} > 0, \qquad \theta = \cos^{-1}\left(-\alpha/\sqrt{4\beta}\right) \in (0, \pi).$$

Substituting (1.53) into the general solution (1.52) yields

$$u = c_1 r^n e^{in\theta} + c_2 r^n e^{-in\theta} = r^n\left[\tilde{c}_1 \cos(n\theta) + \tilde{c}_2 \sin(n\theta)\right], \tag{1.54}$$

where $\tilde{c}_1, \tilde{c}_2$ are arbitrary real constants.

Example 1.14 Show that there is no solution of

$$u_2 - 4u_1 + 8u = 0, \qquad u(0) = 0, \quad u(1000) = 1.$$

Solution: The auxiliary equation is

$$\lambda^2 - 4\lambda + 8 = 0,$$

whose roots are $\lambda_1 = 2(1 + i)$ and $\lambda_2 = 2(1 - i)$. Thus $r = 2^{3/2}$, $\theta = \pi/4$, and so the general solution of the OΔE is

$$u = 2^{3n/2}\left[\tilde{c}_1 \cos(n\pi/4) + \tilde{c}_2 \sin(n\pi/4)\right].$$

Therefore $u(0) = \tilde{c}_1$ and $u(1000) = 2^{1500}\tilde{c}_1$. Consequently, it is not possible to satisfy both conditions simultaneously. This is an example of a *two-point boundary-value problem*. Just as for ODEs, such problems may have a unique solution, an infinite family of solutions, or no solutions, depending on the boundary data. For instance, if the second boundary condition is replaced by $u(1000) = 0$, there is now an infinite family of solutions,

$$u = 2^{3n/2}\tilde{c}_2 \sin(n\pi/4).$$

Alternatively, if the second boundary condition is replaced by $u(1001) = 1$, there is a unique solution

$$u = 2^{3n/2 - 1501} \sin(n\pi/4). \qquad \blacktriangle$$

Case III: A repeated real root. If $\alpha^2 - 4\beta = 0$, the auxiliary equation has only one root, namely $\lambda = -\alpha/2$, so the general solution of the OΔE is not a linear superposition of two independent solutions of the form (1.49). However, we can use reduction of order to solve (1.50), which amounts to

$$u_2 + \alpha u_1 + (\alpha^2/4)\, u = 0. \tag{1.55}$$

As $\alpha \neq 0$, the known solution $u = (-\alpha/2)^n$ can be used for reduction of order. Writing

$$u = (-\alpha/2)^n\, v, \qquad w = v_1 - v,$$

we obtain

$$w_1 - w = 0.$$

Hence

$$v_1 - v = w = c_1,$$

and therefore

$$u = (-\alpha/2)^n\, v = (-\alpha/2)^n\, (c_1 n + c_2). \tag{1.56}$$

Example 1.15 Solve the initial-value problem

$$u_2 - 4u_1 + 4u = 0, \qquad u(0) = 1, \ u(1) = 0.$$

Solution: The auxiliary equation, $\lambda^2 - 4\lambda + 4 = 0$, has only one root, $\lambda = 2$, so the general solution of the OΔE is

$$u = 2^n(c_1 n + c_2).$$

The initial conditions are satisfied when $c_1 = -1$ and $c_2 = 1$.   ▲

1.6 Factorization

Sometimes it is convenient to regard the problem of solving a linear OΔE as the problem of finding the kernel of a linear operator. This approach is widely used for differential equations, for which the operator is a differential operator.

Any OΔE can be written in terms of the *forward shift operator* S_n and the *identity operator* I, which are defined as follows:

$$S_n : n \mapsto n + 1, \qquad I : n \mapsto n, \qquad \text{for all } n \in \mathbb{Z}. \tag{1.57}$$

The identity operator maps every function of n to itself. The forward shift operator replaces n by $n + 1$ in all functions of n, so that

$$S_n\{f(n)\} = f(n + 1), \qquad S_n u = u_1, \qquad S_n u_i = u_{i+1}. \tag{1.58}$$

(The last two equations in (1.58) can be combined by using u_0 as an alternative notation for u. Henceforth, we will use this alternative whenever it is convenient.) Formally, the forward difference operator is the difference between the forward shift and identity operators, as follows:

$$\Delta_n = S_n - I. \tag{1.59}$$

By applying S_n and its inverse repeatedly, we obtain

$$S_n^r\{f(n)\} = f(n + r), \qquad S_n^r u_i = u_{i+r},$$

for each integer r such that $f(n + r)$ (resp. u_r) is defined; in particular, $I = S_n^0$. Care must be taken that S_n is not used to go outside the domain of definition of functions. For instance, $S_n(1/n)$ is undefined at $n = -1$, even though the function $1/n$ is defined there.

The forward shift operator satisfies a simple *product rule*:

$$S_n\{f(n) g(n)\} = f(n + 1) g(n + 1) = S_n\{f(n)\} S_n\{g(n)\}. \tag{1.60}$$

Consequently,

$$S_n^r\{f(n) g(n)\} = f(n + r) g(n + r) = S_n^r\{f(n)\} S_n^r\{g(n)\}, \tag{1.61}$$

for every $r \in \mathbb{Z}$ for which both sides of (1.61) are defined. The product rule (1.60) is significantly simpler than the modified Leibniz rule (1.16), so we shall treat S_n as the fundamental operator from here on. There is another good reason for doing so: when a difference operator is expressed as a linear combination of powers of S_n, its order is obvious. The same need not be true when the same operator is expressed in terms of Δ_n (see Exercise 1.4).

Any first-order linear homogeneous OΔE (1.25) may be written in operator notation as

$$(S_n + a(n) I) u = 0.$$

Therefore the general solution of the OΔE is the set of all functions u that are in the kernel of the operator $S_n + a(n) I$. Similarly, the second-order linear homogeneous OΔE

$$u_2 + a_1(n) u_1 + a_0(n) u = 0, \qquad a_0(n) \neq 0, \tag{1.62}$$

can be written as

$$(S_n^2 + a_1(n) S_n + a_0(n) I) u = 0. \tag{1.63}$$

The method of *factorization* splits the second-order operator in (1.63) into two first-order linear operators which are used successively. This reduces the problem of solving (1.62) to the problem of solving two first-order linear OΔEs.

As an example, consider the second-order constant-coefficient OΔE (1.50) in operator notation:

$$(S_n^2 + \alpha S_n + \beta I) u = 0, \qquad \beta \neq 0. \tag{1.64}$$

If λ_1 and λ_2 are the roots of the auxiliary equation then $\alpha = -(\lambda_1 + \lambda_2)$ and $\beta = \lambda_1 \lambda_2$, so the OΔE can be factorized as follows:

$$(S_n - \lambda_1 I)(S_n - \lambda_2 I) u = 0. \tag{1.65}$$

This may be regarded as a first-order OΔE for $(S_n - \lambda_2 I) u$, whose general solution is

$$(S_n - \lambda_2 I) u = c_1 \lambda_1^n. \tag{1.66}$$

We now solve (1.66) by the integrating factor method; dividing it by λ_2^{n+1} gives

$$\Delta_n(\lambda_2^{-n} u) = \frac{c_1}{\lambda_2} \left(\frac{\lambda_1}{\lambda_2} \right)^n.$$

Therefore the general solution of (1.64) is

$$u = \begin{cases} \tilde{c}_1 \lambda_1^n + c_2 \lambda_2^n, & \lambda_2 \neq \lambda_1; \\ (\tilde{c}_1 n + c_2) \lambda_1^n, & \lambda_2 = \lambda_1. \end{cases} \tag{1.67}$$

Note that the method of factorization does not need to be modified when the auxiliary equation has a repeated root.

It is very convenient to be able to solve a large class of OΔEs by finding the roots of an algebraic equation. This approach works for constant-coefficient OΔEs because $S_n(\lambda^n)$ is a constant multiple of λ^n for each $\lambda \in \mathbb{R}$. The same idea extends to any second-order OΔE that can be factorized as follows:

$$\big(f(n) S_n + \{g(n) - \lambda_1\} I \big) \big(f(n) S_n + \{g(n) - \lambda_2\} I \big) u = 0. \tag{1.68}$$

The first step is to determine the eigenfunctions of the difference operator $f(n) S_n + g(n) I$; these are the nonzero solutions $\varphi(n; \lambda)$ of

$$(f(n) S_n + g(n) I) \varphi(n; \lambda) = \lambda \varphi(n; \lambda). \tag{1.69}$$

The OΔE (1.68) has nonzero solutions of the form $u = \varphi(n; \lambda)$ provided that the auxiliary equation,

$$(\lambda - \lambda_1)(\lambda - \lambda_2) = 0,$$

is satisfied. So if $\lambda_2 \neq \lambda_1$, the general solution of (1.68) is

$$u = c_1 \varphi(n; \lambda_1) + c_2 \varphi(n; \lambda_2). \tag{1.70}$$

If $\lambda_2 = \lambda_1$ then

$$u = \left\{ c_1 \sigma_k \left(\frac{1}{\lambda_1 - g(k)} ; n_0, n \right) + c_2 \right\} \varphi(n; \lambda_1). \qquad (1.71)$$

(The proof of this result is left as an exercise.)

So far, we have used a constructive approach to obtain the class of OΔEs (1.68) that can be solved by finding roots of an auxiliary equation. In practice, however, we will need to recognize such OΔEs when their factors are expanded as follows:

$$f(n)f(n+1)u_2 + f(n)\{g(n+1) + g(n) - \lambda_1 - \lambda_2\}u_1$$
$$+ (g(n) - \lambda_1)(g(n) - \lambda_2)u = 0. \qquad (1.72)$$

The standard form of this class of OΔEs is

$$u_2 + \frac{g(n+1) + g(n) - \lambda_1 - \lambda_2}{f(n+1)} u_1 + \frac{(g(n) - \lambda_1)(g(n) - \lambda_2)}{f(n)f(n+1)} u = 0. \qquad (1.73)$$

For some OΔEs, it is easy to spot suitable functions $f(n)$ and $g(n)$.

Example 1.16 *Cauchy–Euler* difference equations are OΔEs of the form

$$n(n+1)u_2 + \alpha n u_1 + \beta u = 0, \qquad n \geq 1, \quad \beta \neq 0. \qquad (1.74)$$

They belong to the class (1.72), with $f(n) = n$ and $g(n) = 0$. Therefore we seek solutions that satisfy (1.69), that is

$$n\varphi(n+1; \lambda) = \lambda \varphi(n; \lambda), \qquad n \geq 1. \qquad (1.75)$$

Every solution of (1.75) is a constant multiple of

$$\varphi(n; \lambda) = \frac{\lambda^{n-1}}{(n-1)!} . \qquad (1.76)$$

Therefore we can solve the Cauchy–Euler difference equation (1.74) by looking for solutions of the form (1.76), which yields the auxiliary equation

$$\lambda^2 + \alpha\lambda + \beta = 0.$$

If the roots λ_1 and λ_2 of the auxiliary equation are distinct, the general solution of (1.74) is

$$u = \frac{c_1 \lambda_1^{n-1} + c_2 \lambda_2^{n-1}}{(n-1)!} ;$$

if $\lambda_2 = \lambda_1$ then

$$u = \frac{(c_1 n + c_2)\lambda_1^{n-1}}{(n-1)!} . \qquad \blacktriangle$$

Cauchy–Euler equations belong to the class of OΔEs (1.72) for which $g(n)$ is constant. Any OΔE in this class,

$$f(n)f(n+1)u_2 + \alpha f(n)u_1 + \beta u = 0, \qquad \beta \neq 0,$$

reduces to a constant-coefficient OΔE for

$$v = \pi_l\left(f(l); n_0, n\right) u.$$

However, OΔEs of the form (1.72) with non-constant $g(n)$ cannot be solved by this simple substitution; they require the full power of the method of factorization.

Example 1.17   Find the general solution of the OΔE

$$n(n+1)u_2 + 2n(n+1)u_1 + (n+1/2)^2 u = 0, \quad n \geq 1. \tag{1.77}$$

Solution: Clearly, this OΔE is of the form (1.72), with $f(n) = g(n) = n$ and $\lambda_1 = \lambda_2 = -1/2$. The first-order OΔE (1.69) amounts to

$$n\varphi(n+1; -1/2) + (n+1/2)\,\varphi(n; -1/2) = 0, \tag{1.78}$$

so every solution of (1.78) is a multiple of

$$\varphi(n; -1/2) = \prod_{l=1}^{n-1}\left\{-\left(\frac{l+1/2}{l}\right)\right\} = \frac{(-1)^{n-1}\Gamma(n+1/2)}{\Gamma(3/2)(n-1)!}.$$

So, from (1.71), the general solution of (1.77) is

$$u = \left\{c_2 - c_1\sum_{k=1}^{n-1}\frac{2}{2k+1}\right\}\frac{(-1)^{n-1}\Gamma(n+1/2)}{\Gamma(3/2)(n-1)!}$$

$$= \left\{c_2 + c_1(H_{n-1} - 2H_{2n-1} + 2)\right\}\frac{(-1)^{n-1}\Gamma(n+1/2)}{\Gamma(3/2)(n-1)!}. \qquad \blacktriangle$$

One difficulty with the method of factorization is that the OΔE may not be presented in a form whose factorization is obvious. For instance, the OΔE (1.77) in the previous example could have been written in standard form as

$$u_2 + 2u_1 + \left\{1 + \frac{1}{4n(n+1)}\right\}u = 0. \tag{1.79}$$

It is reasonable to ask whether there is a way to factorize (1.79) without resorting to trial-and-error. Therefore let us examine the conditions under which

$$u_2 + a_1(n)u_1 + a_0(n)u = 0, \qquad a_0(n) \neq 0, \tag{1.80}$$

can be factorized. As this OΔE is in standard form, every factorization must be of the form

$$\left(\frac{1}{f(n+1)} S_n + g(n) I\right) (f(n) S_n + h(n) I) u = 0,$$ (1.81)

for some nonzero functions $f(n)$, $g(n)$ and $h(n)$. Expanding (1.81) yields

$$\left\{ S_n^2 + \left(\frac{h(n+1)}{f(n+1)} + f(n) g(n)\right) S_n + g(n) h(n) I \right\} u = 0,$$

so

$$\frac{h(n+1)}{f(n+1)} + f(n) g(n) = a_1(n), \qquad g(n) h(n) = a_0(n).$$ (1.82)

Clearly, the constraints (1.82) are insufficient to determine all three of the functions f, g and h. It is convenient to set $f(n) = -1$; there is no loss of generality in doing so, because this is equivalent to replacing $g(n)$ and $h(n)$ in (1.82) by $-g(n)/f(n)$ and $-f(n) h(n)$, respectively. Therefore the factorization is

$$(S_n - g(n) I) (S_n - h(n) I) u = 0,$$ (1.83)

where

$$g(n) + h(n+1) = -a_1(n), \qquad g(n) h(n) = a_0(n).$$ (1.84)

By eliminating $g(n)$ from (1.84), we obtain the following nonlinear OΔE for the unknown function $h = h(n)$:

$$h_1 h + a_1(n) h + a_0(n) = 0, \qquad a_0(n) \neq 0.$$ (1.85)

This type of OΔE is called a *Riccati equation*. So far, we have reduced the problem of solving (1.80) to the problem of finding a single solution of (1.85). (The general solution depends on an arbitrary constant, so many different factorizations are possible – see Exercise 1.15.) However, for most coefficients $a_1(n)$ and $a_0(n)$, there is no systematic way of finding such a solution. Indeed, equations of the form (1.85) are commonly linearized by the substitution $h = \tilde{u}_1/\tilde{u}$, which converts (1.85) into the original OΔE (1.80) with $\tilde{u}$ replacing u. Yet this is the OΔE that we are trying to solve!

If $a_1(n) = 0$, the Riccati equation (1.85) may be solved formally by taking logarithms to obtain the linear first-order OΔE

$$(S_n + I) \ln(h) = \ln(-a_0(n)),$$

which can be solved by the integrating factor method. Care may be needed when arbitrary constants are chosen, to ensure that the solution of (1.80) is real-valued. There is a more direct approach to the problem of solving (1.80) when $a_1(n) = 0$, which is discussed in Exercise 1.16. If $a_1(n)$ is nonzero, one must hope to be able to spot a solution of (1.85). This may or may not be easier than spotting a nonzero solution of (1.80).

Example 1.18   The Riccati equation for the OΔE (1.79) is

$$h_1 h + 2h + \left\{ 1 + \frac{1}{4n(n+1)} \right\} = 0, \tag{1.86}$$

which suggests that there may be a solution of the form

$$h = \frac{\alpha n + \beta}{n}, \tag{1.87}$$

where α and β are constants. Substituting (1.87) into (1.86) leads to

$$((\alpha + 2)n + \alpha + \beta + 2)(\alpha n + \beta) + (n + 1/2)^2 = 0,$$

which must hold for each $n \geq 1$. Hence there are three constraints on α and β, which are obtained by comparing terms that are multiplied by each power of n, as follows.

$$\begin{aligned}
\text{Terms multiplied by } n^2 : \quad & \alpha(\alpha + 2) + 1 = 0, \\
\text{Terms multiplied by } n : \quad & \alpha(\alpha + \beta + 2) + \beta(\alpha + 2) + 1 = 0, \\
\text{Remaining terms} : \quad & \beta(\alpha + \beta + 2) + 1/4 = 0.
\end{aligned}$$

There is a unique solution, $(\alpha, \beta) = (-1, -1/2)$, so

$$h = -(1 + 1/(2n)).$$

At this stage, $g(n)$ can be found from (1.84), which produces the factorization (1.83) as follows:

$$\left(S_n + \frac{n + 1/2}{n + 1} I \right) \left(S_n + \frac{n + 1/2}{n} I \right) u = 0.$$

Although this is not the same as the factorization (1.68) that is used in Example 1.17, it is equally effective.   ▲

1.7 Transformation to a known linear OΔE

So far, we have tried to find a nonzero solution of a given second-order linear homogeneous OΔE by two approaches.

Inspection: In practice, inspection usually amounts to seeking a solution of the linear OΔE or the associated Riccati equation by using an *ansatz*, that is, a prior assumption about the form of the solution. The most commonly used assumptions are:

(i) that u is a low-order polynomial in n;

(ii) that u is the ratio of two low-order polynomials in n.

We have used (i) in Example 1.12 and (ii) in Example 1.18. Inspection may also be used to find a factorization of a difference operator directly, without reference to the Riccati equation.

Compiling a list of solvable problems: We have already encountered a wide range of second-order OΔEs that can be solved. It is useful to list these, by putting each OΔE in standard form and then writing down the coefficients $a_1(n)$ and $a_0(n)$, including parameters or arbitrary functions where appropriate.

(i) OΔEs with constant coefficients have $a_1(n) = \alpha$ and $a_0(n) = \beta$, where $\beta \neq 0$.

(ii) Arbitrary functions and parameters occur in the coefficients of the large class of problems (1.73) that can be solved by finding the roots of an algebraic equation.

(iii) There is another rich source of solvable second-order linear OΔEs that we have not yet considered, namely three-term recurrence relations for special functions and orthogonal polynomials. Some of these are listed in Table 1.2. For most of these OΔEs, we have listed two linearly independent solutions; where only one solution has been listed, reduction of order will give the general solution in terms of an unevaluated sum.

(iv) For OΔEs with $a_1(n) = 0$, the Riccati equation can always be linearized, so these OΔEs can also be solved.

(v) Finally, we have met a few OΔEs that can be solved by inspection followed by reduction of order; these can be added to the list.

It may seem that, if we can neither recognize a given second-order linear homogeneous OΔE nor find a nonzero solution by inspection, we have no hope of obtaining the general solution. However, the list above is made far more useful by a simple observation: any given linear homogeneous OΔE can be transformed into a whole class of such OΔEs. This class is constructed by using three kinds of transformation, each of which maps the set of all p^{th}-order linear homogeneous OΔEs to itself. By applying some or all of these transformations to a given OΔE, one may be able to reduce it to a known OΔE. For simplicity, we restrict attention to second-order OΔEs,

$$u_2 + a_1(n)\, u_1 + a_0(n)\, u = 0, \qquad a_0(n)\, a_1(n) \neq 0, \qquad (1.88)$$

where $D = \mathbb{Z}$. The transformations can equally well be applied to higher-order OΔEs. If $D \neq \mathbb{Z}$, transformations that change n will change the domain, so

Table 1.2 *Coefficients and stretch invariants for functions that solve* $u_2 + a_1(n)u_1 + a_0(n)u = 0$. *(Adapted from Abramowitz and Stegun (1965).)*

Class of functions	Independent solutions	$a_1(n)$	$a_0(n)$	Stretch invariant $\rho(n)$
Bessel functions	$J_n(x)$, $Y_n(x)$	$-\dfrac{2(n+1)}{x}$	1	$\dfrac{4n(n+1)}{x^2}$
Modified Bessel functions	$I_n(x)$, $(-1)^n K_n(x)$	$\dfrac{2(n+1)}{x}$	-1	$-\dfrac{4n(n+1)}{x^2}$
Spherical Bessel functions	$j_n(x)$, $y_n(x)$	$-\dfrac{2n+3}{x}$	1	$\dfrac{(2n+3)(2n+1)}{x^2}$
Parabolic cylinder functions	$U(n,x)$, $\Gamma(\tfrac12-n)V(n,x)$	$\dfrac{2x}{2n+3}$	$-\dfrac{2}{2n+3}$	$-\dfrac{2x^2}{2n+1}$
Parabolic cylinder functions	$V(n,x)$, $U(n,x)/\Gamma(\tfrac12-n)$	$-x$	$-(n+\tfrac12)$	$-\dfrac{2x^2}{2n+1}$
Associated Legendre functions	$P_n^\mu(x)$, $Q_n^\mu(x)$	$-\dfrac{(2n+3)x}{n+2-\mu}$	$\dfrac{n+1+\mu}{n+2-\mu}$	$\dfrac{(2n+3)(2n+1)x^2}{(n+1)^2-\mu^2}$
Associated Legendre functions	$P_\mu^n(x)$, $Q_\mu^n(x)$, $\quad x\neq\pm1$	$\dfrac{2(n+1)x}{\sqrt{x^2-1}}$	$(n-\mu)(n+1+\mu)$	$\dfrac{4n(n+1)x^2}{(x^2-1)(n-\mu)(n+1+\mu)}$
Legendre polynomials	$P_n(x)$	$-\dfrac{(2n+3)x}{n+2}$	$\dfrac{n+1}{n+2}$	$\dfrac{(2n+3)(2n+1)x^2}{(n+1)^2}$
Hermite polynomials	$H_n(x)$	$-2x$	$2(n+1)$	$\dfrac{2x^2}{n+1}$
Generalized Laguerre polynomials	$L_n^{(\alpha)}(x)$, $\quad \alpha>-1$	$\dfrac{x-(2n+3+\alpha)}{n+2}$	$\dfrac{n+1+\alpha}{n+2}$	$\dfrac{(x-(2n+3+\alpha))(x-(2n+1+\alpha))}{(n+1)(n+1+\alpha)}$

care is needed to ensure that the transformed OΔE is solved on the correctly transformed domain.

The idea of transformation is to introduce new variables $\hat{n}$ and $\hat{u}$ that enable us to write the OΔE (1.88) as another second-order OΔE,

$$\hat{u}_2 + \tilde{a}_1(\hat{n})\,\hat{u}_1 + \tilde{a}_0(\hat{n})\,\hat{u} = 0, \qquad \tilde{a}_0(n)\,\tilde{a}_1(n) \neq 0. \qquad (1.89)$$

If the new variables depend only on n and u, they are determined by a *fibre-preserving transformation*; we drop the qualifier 'fibre-preserving' for now. Roughly speaking[3], such a transformation is an invertible mapping,

$$\Psi : (n, u) \mapsto (\hat{n}, \hat{u}) = (v(n), \psi_n(u)) \qquad (1.90)$$

(where each ψ_n is a smooth function), that preserves the order of the OΔE.

For now, suppose that $\psi_n(u) = u$ for every n. Replacing n by $n+i$ in (1.90) and treating $\hat{n} = v(n)$ as the new independent variable, we obtain

$$\Psi : u_i \mapsto S_n^i\big(u(n)\big) = S_n^i\big(\hat{u}(v(n))\big) = \hat{u}(v(n + i)).$$

The original OΔE involves the values of u at three consecutive points: $n + 1$ is adjacent to n and $n + 2$ for every $n \in \mathbb{Z}$. The transformed OΔE will be second-order only if $v(n + 1)$ is adjacent to $v(n)$ and $v(n + 2)$ for each $n \in \mathbb{Z}$. Thus, the only valid transformations of the independent variable are translations and reflections. For such transformations, using $\hat{u}_{\hat{i}}$ as shorthand for $\hat{u}(\hat{n} + \hat{i})$,

$$\Psi : u_i \mapsto \hat{u}_{\hat{i}}, \quad \text{where} \quad \hat{i} = v(n + i) - v(n). \qquad (1.91)$$

As adjacency is preserved, $\hat{i}$ depends only on i.

A *translation* maps each $n \in \mathbb{Z}$ to $\hat{n} = n + m_0$, where m_0 is a fixed integer. Consequently, $v(n + i) - v(n) = i$, so the OΔE (1.88) can be rewritten as

$$\hat{u}_2 + a_1(\hat{n} - m_0)\,\hat{u}_1 + a_0(\hat{n} - m_0)\,\hat{u} = 0. \qquad (1.92)$$

This is of the form (1.89), with

$$\tilde{a}_1(\hat{n}) = a_1(\hat{n} - m_0), \qquad \tilde{a}_0(\hat{n}) = a_0(\hat{n} - m_0). \qquad (1.93)$$

If $\hat{u} = F(\hat{n})$ is a solution of the transformed OΔE (1.92), the corresponding solution of (1.88) is $u = F(n + m_0)$.

Example 1.19 Use a translation to find the general solution of

$$u_2 + nu_1 - u = 0.$$

Solution: Here $a_1(n) = n$ and $a_0(n) = -1$, so (1.93) gives

$$\tilde{a}_1(\hat{n}) = \hat{n} - m_0 \quad \text{and} \quad \tilde{a}_0(\hat{n}) = -1.$$

---

[3] A precise definition is given in Chapter 4.

Comparison with Table 1.2 shows that by choosing $m_0 = -1$ and $x = 2$, we can obtain the general solution of (1.92) in terms of modified Bessel functions:

$$\hat{u} = c_1 I_{\hat{n}}(2) + c_2(-1)^{\hat{n}} K_{\hat{n}}(2).$$

Therefore the general solution of the original O∆E is

$$u = c_1 I_{n-1}(2) + c_2(-1)^{n-1} K_{n-1}(2). \qquad\qquad ▲$$

Now let us apply the same approach to examine the effect on the O∆E of a *reflection*. This maps n to $\hat{n} = m_0 - n$, where m_0 is a given integer, and so $v(n + i) - v(n) = -i$. Therefore the O∆E (1.88) amounts to

$$\hat{u}_{-2} + a_1(m_0 - \hat{n})\,\hat{u}_{-1} + a_0(m_0 - \hat{n})\,\hat{u} = 0.$$

This can be put into standard form by replacing $\hat{n}$ by $\hat{n} + 2$ and rearranging, as follows:

$$\hat{u}_2 + \frac{a_1(m_0 - 2 - \hat{n})}{a_0(m_0 - 2 - \hat{n})}\,\hat{u}_1 + \frac{1}{a_0(m_0 - 2 - \hat{n})}\,\hat{u} = 0. \qquad (1.94)$$

Consequently, the images of the coefficient functions a_1 and a_2 are

$$\tilde{a}_1(\hat{n}) = \frac{a_1(m_0 - 2 - \hat{n})}{a_0(m_0 - 2 - \hat{n})}, \qquad \tilde{a}_0(\hat{n}) = \frac{1}{a_0(m_0 - 2 - \hat{n})}. \qquad (1.95)$$

The above reflection amounts to a reflection about $n = 0$ followed by a translation by m_0. It is convenient to retain the parameter m_0, however, so that all reflections can be examined at the same time.

So far we have focused on transformations of the independent variable. All other fibre-preserving transformations are obtained by composing these with transformations that do not affect the independent variable. To map (1.88) to an O∆E of the form (1.89), such a transformation must be linear in u. Moreover, every O∆E that can be derived in this way may be obtained from a *stretch*[4],

$$(\hat{n}, \hat{u}) = (n, h(n)u), \qquad h(n) \neq 0, \qquad (1.96)$$

which transforms the coefficients $a_i(n)$ as follows:

$$\tilde{a}_1(n) = \frac{h(n + 2)}{h(n + 1)}\, a_1(n), \qquad \tilde{a}_0(n) = \frac{h(n + 2)}{h(n)}\, a_0(n). \qquad (1.97)$$

A function $\rho(n)$ is called an *invariant* of a given transformation if $\tilde{\rho}(n) = \rho(n)$ for all $n \in \mathbb{Z}$. There is an invariant whose form is common to every stretch, namely

$$\rho(n) = \frac{a_1(n - 1)\,a_1(n)}{a_0(n)}. \qquad (1.98)$$

---

[4] These results are proved by differential elimination (see Chapter 2).

So no stretch can change an OΔE (1.88) into another OΔE with a different $\rho(n)$. Consequently, if we aim to simplify a given OΔE by using a stretch, we must first calculate $\rho(n)$ and then find out which (if any) of the known OΔEs share the same $\rho(n)$. Some examples are listed in Table 1.2. If such an OΔE can be found, write its coefficients as $\tilde{a}_1(n)$ and $\tilde{a}_0(n)$. Then the stretch

$$h(n) = \pi_l\left(\frac{\tilde{a}_1(l-1)}{a_1(l-1)} ; n_0, n\right) \tag{1.99}$$

transforms (1.88) into the known OΔE (1.89).

Some stretches have no effect on the coefficients[5], namely *scalings* by a constant real factor: if $h(n) = c \neq 0$ then $\tilde{a}_1(n) = a_1(n)$ and $\tilde{a}_0(n) = a_0(n)$. Similarly, linear superposition does not change any coefficients, which is why we have no need to consider such transformations at present.

Typically, $\rho(n)$ is not invariant under translations or reflections. However, in each case $\rho(n)$ transforms in a simple way, which makes it easy to find all other functions $\tilde{\rho}(n)$ that can be obtained from it. The translation $v(n) = n + m_0$ gives

$$\tilde{\rho}(n) = \frac{a_1(n - m_0 - 1)\,a_1(n - m_0)}{a_0(n - m_0)} = \rho(n - m_0).$$

Similarly, each reflection $v(n) = m_0 - n$ produces

$$\tilde{\rho}(n) = \frac{a_1(m_0 - 2 - n)\,a_1(m_0 - 1 - n)}{a_0(m_0 - 1 - n)} = \rho(m_0 - 1 - n).$$

Hence, if $\rho(n)$ for the given OΔE can be mapped by a translation or a reflection to the $\rho(n)$ for an OΔE whose solution is known, the given OΔE can be solved.

Example 1.20 Find the general solution of

$$u_2 - u_1 + \frac{4}{4n^2 - 1}\,u = 0. \tag{1.100}$$

Solution: First calculate

$$\rho(n) = \frac{a_1(n-1)\,a_1(n)}{a_0(n)} = n^2 - 1/4 = (n + 1/2)(n - 1/2).$$

There are no special functions in Table 1.2 with this $\rho(n)$; however the spherical Bessel functions $j_n(x), y_n(x)$ have

$$\rho(n) = (2n + 3)(2n + 1)/x^2.$$

By choosing $x = 2$, this reduces to $\rho(n) = (n + 3/2)(n + 1/2)$, which is a translated version of the desired $\rho(n)$. A combination of translation and stretching

---

[5] Transformations that map a difference equation to itself are called *symmetries* of the equation. Perhaps surprisingly, symmetries are generally very useful when one wants to find exact solutions of difference equations. The reason for this is described in the next chapter.

is needed to transform the given OΔE into the OΔE for $j_n(x)$ and $y_n(x)$. This is achieved by translating first to obtain the correct $\rho(n)$, then calculating the appropriate stretch. The translation replaces n by $n - 1$. Applying this to the OΔE for the spherical Bessel functions (with $x = 2$) produces

$$u_2 - (n + 1/2)\, u_1 + u = 0, \tag{1.101}$$

whose general solution is

$$u = c_1 j_{n-1}(2) + c_2 y_{n-1}(2).$$

The translated OΔE (1.101) has $\rho(n) = n^2 - 1/4$; equation (1.99) yields the stretch

$$h(n) = \Gamma(n - 1/2).$$

Therefore the general solution of the given OΔE (1.100) is

$$u = \frac{c_1 j_{n-1}(2) + c_2 y_{n-1}(2)}{\Gamma(n - 1/2)}. \qquad\blacktriangle$$

Notes and further reading

There is a huge range of literature on linear OΔEs and their applications, which include control theory, economics, combinatorics and mathematical biology. To learn more about these, I recommend the excellent texts by Elaydi (2005), Mickens (1990) and Kelley and Peterson (2001). Each one describes a number of methods that are not included in this chapter, including the z-transform (which is a discrete analogue of the Laplace transform), methods for approximating sums, Sturm-Liouville theory, and methods for linear systems of OΔEs. Kelley and Peterson's book also includes a thorough discussion of boundary-value problems for both linear and nonlinear OΔEs. Generating functions provide a useful link between solutions of linear OΔEs and linear ODEs (see Exercise 1.10). They can also greatly simplify the problem of rewriting a given sum in closed form. Wilf (2006) is a very readable introduction to generating functions; the previous edition of this text may be downloaded (free of charge) from the author's website.

Computer algebra is immensely useful for dealing with difference equations of all kinds. In particular, the packages Rsolve in *Mathematica* and rsolve and LREtools in *Maple* provide various tools for solving and manipulating linear (and some nonlinear) OΔEs. It is well worth becoming familiar with at least one of these packages.

Exercises

1.1 Let $f_1(n), ..., f_p(n)$ be solutions of a pth-order linear homogeneous OΔE on a regular domain D. Show that they are linearly independent if and only if there exists $n \in D$ such that

$$\det \begin{bmatrix} f_1(n) & f_2(n) & \cdots & f_p(n) \\ f_1(n+1) & f_2(n+1) & \cdots & f_p(n+1) \\ \vdots & \vdots & & \vdots \\ f_1(n+p-1) & f_2(n+p-1) & \cdots & f_p(n+p-1) \end{bmatrix} \neq 0.$$

The above matrix is called the *Casoratian*; it is analogous to the Wronskian matrix for linear ODEs.

1.2 For each of the following, evaluate $\sum_{k=1}^{n-1} f(k)$ in closed form:

(a) $f(k) = k^4$;
(b) $f(k) = k/\{(k+1)(k+3)(k+4)\}$;
(c) $f(k) = k^2(-1)^k$;
(d) $f(k) = 2^k \cos(2\pi k/3)$.

1.3 Let $A^{\pm}(a; n_0, n_1) = \sum_{k=n_0}^{n_1-1} a^k(-1)^{k(k\pm 1)/2}$. Assuming that $a \neq 1$, evaluate these sums and calculate their limits as $a \to 1$. [Hint: the decompositions

$$1 = \tfrac{1}{2}(1+(-1)^k) + \tfrac{1}{2}(1-(-1)^k), \quad (-1)^k = \tfrac{1}{2}(1+(-1)^k) - \tfrac{1}{2}(1-(-1)^k),$$

are useful.] Compare your answer with Example 1.5.

1.4 Show that $(\Delta_n)^2 u - \Delta_n u - 2u = 0$ is *not* a second-order OΔE. This illustrates one way in which differences do not behave like derivatives.

1.5 The summation technique for $k^{(r)}$ has the following simple extension. We define

$$(k+\alpha)^{(r)} = \frac{\Gamma(k+1+\alpha)}{\Gamma(k+1+\alpha-r)}, \quad r \in \mathbb{Z}, \quad \alpha \in \mathbb{R}/\mathbb{Z}.$$

Show that $\Delta_k((k+\alpha)^{(r)}) = r(k+\alpha)^{(r-1)}$ and evaluate $\sum_{k=1}^{n-1} 1/(4k^2 - 1)$.

1.6 Find the general solution of each of the following:

(a) $u_1 + u = (-1)^n, \qquad n \geq 0$;

(b) $u_1 + (1/n)\, u = 0, \qquad n \leq -1;$

(c) $u_1 + ((1 + n)/(1 - n))\, u = n^3(n + 1), \qquad n \geq 2.$

1.7   Solve the reduced OΔE (1.46) in Example 1.12.

1.8   Solve each of the following initial-value problems:

(a) $u_2 - u_1 - u = 0, \qquad u(0) = 0,\ u(1) = 1$ (Fibonacci);

(b) $u_2 - 2u_1 + 4u = 0, \qquad u(1) = 1,\ u(2) = 0;$

(c) $u_2 + 2u_1 + u = (-1)^n, \qquad u(0) = u(1) = 0;$

(d) $u_3 - 2u_2 + 3u = 0, \qquad u(0) = 0,\ u(1) = 3,\ u(2) = 2.$

1.9   Let $u = f(n)$ be any solution of $u_2 + \alpha u_1 + \beta u = 0,\ n \geq 0$, and define the *exponential generating function* to be

$$G(x) = \sum_{n=0}^{\infty} \frac{x^n}{n!}\, f(n).$$

Show that $y = G(x)$ is a solution of $y'' + \alpha y' + \beta y = 0$. Use the general solution of this ODE to obtain the general solution u of the ODE. [This demonstrates the close link between ODEs and OΔEs with constant coefficients.]

1.10   Some OΔEs with non-constant coefficients can be solved by constructing an ODE for the exponential generating function $G(x)$ or for the *generating function*

$$F(x) = \sum_{n=0}^{\infty} f(n)x^n.$$

Once the ODE has been solved, it is straightforward to recover the function $f(n)$ by evaluating derivatives of $F(x)$ or $G(x)$ at $x = 0$. For example, show that an OΔE of the form

$$(\alpha n + \beta)\, u_2 + (\gamma n + \delta)\, u_1 + (\epsilon n + \eta)\, u = 0, \qquad n \geq 0,$$

where $\alpha, \ldots, \eta$ are constants, can be reduced to a first-order linear ODE for $F(x)$. Hence obtain the general solution of

$$(n + 2)\, u_2 + u_1 - (n + 1)\, u = 0, \qquad n \geq 0.$$

1.11   Any equation of the form

$$u\, u_1 + c(n)\, u_1 + a(n)\, u + b(n) = 0, \qquad b(n) \neq a(n)\, c(n)$$

is called a scalar *Riccati difference equation*. (So far, we have only dealt with Riccati equations for which $c(n) = 0$.) Find a substitution that transforms this equation into a second-order linear homogeneous OΔE. Show that any two linearly independent solutions of the second-order OΔE correspond to two distinct solutions of the Riccati equation.

1.12 Find the general solution of each of the following OΔEs:

(a) $u_2 - \{(2n+3)/(n+2)\} u_1 + \{(n+1)/(n+2)\} u = 1,$ $n \geq 0;$

(b) $u_2 - \{(2n+3)/(n+2)\} u_1 + \{n/(n+1)\} u = 3(n+3),$ $n \geq 1;$

(c) $u_2 - (3+1/n) u_1 + 2(1+1/n) u = 1/n,$ $n \geq 1.$

1.13 Prove the result (1.71).

1.14 Let $u = f(n)$ be a nonzero solution of a second-order linear homogeneous OΔE. Find a factorization of the OΔE. Explain the method of reduction of order in terms of factorization.

1.15 Check that for each real constant c whose modulus is not 1,

$$S_n^2 + I = \left\{ S_n + \left(\frac{(-1)^n - c}{1 + c(-1)^n} \right) I \right\} \left\{ S_n + \left(\frac{(-1)^n + c}{1 - c(-1)^n} \right) I \right\}.$$

Now suppose that, for a given function f,

$$(S_n - f(n) I)^2 = (S_n - v(n) I)(S_n - w(n) I).$$

Write down the system of OΔEs that determines v and w, and find its general solution.

1.16 Show how the general solution of $u_2 + a_0(n) u = 0$ can be obtained from the general solution of the first-order OΔEs

$$\hat{u}_1 + a_0(2\hat{n}) \hat{u} = 0, \qquad \hat{u}_1 + a_0(2\hat{n} + 1) \hat{u} = 0.$$

[Hint: what happens when n is even?] Extend your result to all OΔEs of the form $u_2 = \omega(n, u)$.

1.17 Use the previous question to solve $u_2 - e^{-n} u = 0$. Compare this method with the Riccati equation method: which is simpler for this problem?

1.18 Use one of the transformations from §1.7 to show that

$$u = c_1 n \sum_{k=1}^{-(n+1)} (k-1)! + c_2 n$$

is the general solution of

$$u_2 - u_1 - (1/n) u = 0, \qquad n \leq -3.$$

[Hint: look at Example 1.11.]

1.19 Use the transformations of §1.7 to find the general solution of

$$u_2 - 4u_1 + (1 - 2n) u = 0.$$

Hence solve

$$u_2 - 4u_1 + (1 - 2n) u = 2n^2.$$

1.20* The well-known method of *variation of parameters* can be used to solve a given linear inhomogeneous ODE, provided that the general solution of the associated homogeneous equation is known. Construct an analogue of this method for inhomogeneous second-order linear OΔEs.

2

Simple symmetry methods for OΔEs

Why bother with a cunning plan
when a simple one will do?

(Terry Pratchett, Thud![1])

This chapter develops solution techniques for nonlinear OΔEs, using an idea that is at the heart of most methods for solving nonlinear ODEs. In many instances, one can simplify a given ODE by introducing a change of variables. Over a century ago, the Norwegian mathematician Sophus Lie showed how to make this process systematic by exploiting symmetries of the ODE. It turns out that Lie's approach also works for OΔEs, except that the calculations are done differently. A similar approach is used to construct first integrals of a given OΔE.

2.1 Nonlinear OΔEs: some basics

We begin by extending the basic definitions that we have used for linear OΔEs. A scalar OΔE on a domain $D \subset \mathbb{Z}$ is an equation that can be written as

$$\mathcal{A}(n, u, u_1, \ldots, u_p) = 0, \qquad n \in D, \tag{2.1}$$

where $\mathcal{A}$ is a given function. A point $n \in D$ is a *regular point* of the OΔE (2.1) if $\mathcal{A}$ depends on both u and u_p there; otherwise it is *singular*. Clearly, the OΔE is of order p only at regular points. A regular domain is a set of consecutive regular points.

A *forward* OΔE has the form

$$u_p = \omega(n, u, u_1, \ldots, u_{p-1}), \qquad n \in D. \tag{2.2}$$

A forward OΔE with p initial conditions, $u(n_0), \ldots, u(n_0 + p - 1)$, determines $u(n)$ uniquely for each $n \geq n_0$, provided that $\omega(n, u, \ldots, u_{p-1})$ is finite on this solution for each $n \in D$. For an initial-value problem that breaches this condition, u becomes unbounded or 'blows up' (see Exercise 2.1).

---

[1] Published by Doubleday, 2005.

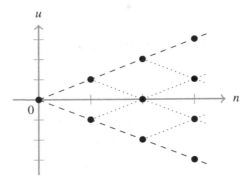

Figure 2.1 Solutions of $u_1 - u = \pm h$, $u(0) = 0$; the set of points $(n, u(n))$ on any path going from left to right is a solution. Only the outer (dashed) paths converge to the smooth solutions (2.6) as $h \to 0$.

Similarly, a *backward* OΔE,

$$u = \Omega(n, u_1, u_2, \ldots, u_p), \qquad n \in D, \tag{2.3}$$

with p conditions, $u(n_0), \ldots, u(n_0 - p + 1)$, has a unique solution, $u(n)$, for each $n \le n_0 - p + 1$, provided that Ω is finite on the whole solution. Consequently, the general solution of a typical[2] forward or backward p^{th}-order OΔE on a regular domain depends on p arbitrary constants. This result does not apply to nonlinear OΔEs that cannot be put into either forward or backward form (at least, in principle[3]), as the following example shows.

Example 2.1 The forward Euler discretization of the initial-value problem

$$(\mathrm{d}y/\mathrm{d}x)^2 = 1, \qquad x \ge 0, \qquad y(0) = 0, \tag{2.4}$$

with constant step length $h > 0$, is

$$((u_1 - u)/h)^2 = 1, \qquad n \ge 0, \qquad u(0) = 0; \tag{2.5}$$

here, u is the approximation to $y(nh)$. The problem (2.4) has two solutions:

$$y(x) = x, \qquad y(x) = -x. \tag{2.6}$$

These are the only solutions, as $y(x)$ is differentiable for all $x \ge 0$. By contrast, solutions of difference equations are not even continuous. So (2.5) amounts to

$$u_1 = u \pm h, \qquad n \ge 0,$$

---

[2] Here 'typical' means that a solution exists for almost all initial conditions. Exercise 2.1(b) deals with an OΔE (on a regular domain) that looks innocuous, but has no solutions.

[3] For example, $u_1 + \exp(u_1) = u + \exp(2u)$ could, in principle, be rewritten in forward or backward form, because $x + \exp(kx)$ is a continuous, strictly monotonic function of $x \in \mathbb{R}$ for each $k \ge 0$; however, neither ω nor Ω would be in closed form.

where the choice of sign is arbitrary at each step. This leads to an infinite family of solutions that satisfy the initial condition $u = 0$ (see Figure 2.1), including:

$$u = nh,$$
$$u = (1 - (-1)^n)h/2,$$
$$u = 2h \, | \sin(n\pi/4)| \sin(n\pi/4).$$

Only the first of these is a discretization of a solution of (2.4). ▲

These observations generalize to all OΔEs that cannot (in principle) be written in forward or backward form. At each step, one has a free choice of which of the various possible values of u to pick. (This is not so for ODEs, whose solutions must be differentiable.) Thus, there are many solutions corresponding to each choice of initial conditions.

To avoid such complications, we consider only OΔEs that are in either forward or backward form, with a regular domain that is appropriate to the form. If the domain is $\mathbb{Z}$, we require that the OΔE has both a forward and a backward form. Then any set of p consecutive values that does not cause blow-up, $\{u(n_0), \ldots, u(n+p-1)\}$, determines a unique solution on the whole domain. We also restrict attention to problems for which ω (resp. Ω) depends smoothly on its continuous arguments (which are u and each u_i) almost everywhere.

For forward OΔEs, D is a regular domain if

$$\omega_{,1}(n, u, \ldots, u_{p-1}) \neq 0, \qquad n \in D.$$

We use the subscript $, k$ to denote the partial derivative of a given function with respect to its kth continuous argument (ignoring any discrete arguments). For instance, $\omega_{,1}$ and $\omega_{,k+1}$ stand for $\partial\omega/\partial u$ and $\partial\omega/\partial u_k$, respectively[4].

Similarly, D is a regular domain for the backward equation (2.3) if $\Omega_{,p} \neq 0$ whenever $n \in D$. Any backward OΔE can be mapped to a forward OΔE by the reflection $n \mapsto p - n$, so we will usually discuss forward OΔEs, omitting the qualifier 'forward' unless it is essential.

2.2 What is a symmetry?

2.2.1 Symmetries of geometrical objects

A symmetry of a geometrical object is an invertible transformation that maps the object to itself. Individual points of an object may be mapped to different points, but the object as a whole is unchanged by any symmetry. For instance,

---

[4] It seems more natural to use ω_k for $\partial\omega/\partial u_k$, but the advantage of the above notation is that it can also be used without modification for partial difference equations.

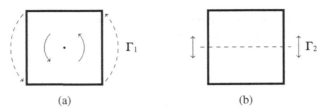

Figure 2.2 Some symmetries of a square: (a) rotation by $\pi/2$; (b) reflection.

consider the rotations of a square about its centre (see Figure 2.2). The square is mapped to itself if the angle of rotation is an integer multiple of $\pi/2$.

If the object has some structure associated with it, every symmetry must preserve this structure (for otherwise the object would be changed). For instance, if the square is rigid, no part of it can be stretched or squashed; the distance between each pair of points of the square must be preserved. All distance-preserving transformations of Euclidean space are composed of translations, rotations and reflections.

Every object has at least one symmetry, namely the identity map (id) that maps each point of the object to itself. (As id leaves each point unchanged, the object as a whole is unchanged.) Suppose that Γ_1 and Γ_2 are symmetries of an object. When they are applied in turn, first Γ_1, then Γ_2, the composite transformation $\Gamma_2\Gamma_1$ is a symmetry, because it leaves the object unchanged and is invertible. Similarly $\Gamma_1\Gamma_2$ is a symmetry, although it need not be the same symmetry as $\Gamma_2\Gamma_1$. The inverse of a symmetry Γ is the transformation Γ^{-1} that reverts the object to its original state, so that $\Gamma^{-1}\Gamma = \text{id}$. Clearly, Γ^{-1} is itself a symmetry whose inverse is $(\Gamma^{-1})^{-1} = \Gamma$, so $\Gamma\Gamma^{-1} = \text{id}$.

The above properties of symmetries (and the fact that composition of transformations is associative) imply that the set $\mathcal{G}$ of all symmetries of a geometrical object is a group[5]. If $\mathcal{G}$ contains finitely many distinct symmetries, it is a finite group; otherwise it is an infinite group.

It is useful to be able to write down $\mathcal{G}$ in an economical way. This can be achieved by using group generators and relations. The symmetries $\Gamma_1, \ldots, \Gamma_R$ are *generators* of $\mathcal{G}$ if one can write every symmetry as a product of some of the generators Γ_i and their inverses. *Relations* are constraints on the products of the generators; a sufficient number of relations must be given to allow $\mathcal{G}$ to be reconstructed from the generators and relations.

---

[5] A *group* is a set, G, that is closed under an associative operation (called group multiplication) and has the following properties:

   (i) G contains an identity element e such that $ge = g = eg$ for every $g \in G$;

  (ii) for each $g \in G$ there exists an inverse $g^{-1} \in G$ such that $gg^{-1} = e = g^{-1}g$.

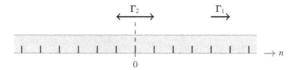

Figure 2.3 Some symmetries of a (rigid) ruler that is marked at every integer. $\Gamma_1 : n \mapsto n + 1$ moves the ruler one mark to the right; $\Gamma_2 : n \mapsto -n$ reflects (flips) the ruler about the mark at 0.

Example 2.2   The group of symmetries of the square is called the dihedral group D_4. Figure 2.2 shows the following generators of the group:

Γ_1 : rotation by $\pi/2$ (anticlockwise) about the square's centre;

Γ_2 : reflection in a centreline.

These generators are subject to the relations

$$(\Gamma_1)^4 = (\Gamma_2)^2 = (\Gamma_1\Gamma_2)^2 = \text{id}.$$

It is easy to deduce that the group has eight distinct elements,

$$D_4 = \left\{ \text{id}, \Gamma_1, (\Gamma_1)^2, (\Gamma_1)^3, \Gamma_2, \Gamma_1\Gamma_2, (\Gamma_1)^2\Gamma_2, (\Gamma_1)^3\Gamma_2 \right\},$$

and that $\Gamma_1\Gamma_2 \neq \Gamma_2\Gamma_1$.   ▲

Example 2.3   In §1.7, we considered the transformations of the set of integers $\mathbb{Z}$ that preserve adjacency. If we treat $\mathbb{Z}$ as a set of equally spaced indistinguishable marks on a rigid, infinitely long ruler, these transformations are precisely the symmetries of the marked ruler (see Figure 2.3). In terms of the usual co-ordinate n, the group of such transformations is generated by

$$\Gamma_1 : n \mapsto n + 1 \qquad \text{(translation by one mark)},$$
$$\Gamma_2 : n \mapsto -n \qquad \text{(reflection about } n = 0\text{)},$$

subject to the relations

$$(\Gamma_2)^2 = (\Gamma_1\Gamma_2)^2 = \text{id}.$$

In this case, $\mathcal{G}$ is the (countably) infinite dihedral group D_∞. It is worth asking whether the rigid structure is really necessary. What happens if the ruler is continuously deformable, so that it can stretch and contract, but not break? Once again, each symmetry of $\mathbb{Z}$ must map the set of all marks, which represents $\mathbb{Z}$, to itself. As the map is continuous, neighbouring points in $\mathbb{Z}$ must remain neighbours, so the freedom to stretch is irrelevant to the marked points. Therefore, in effect, all symmetries of $\mathbb{Z}$ preserve a rigid structure.   ▲

Some objects have symmetry groups that are not generated by a finite number of symmetries. For instance, the symmetries of the Euclidean real line $\mathbb{R}$ include every translation

$$\Gamma_\varepsilon : x \mapsto x + \varepsilon, \tag{2.7}$$

where ε is a fixed real number. As Γ_ε is a symmetry for each $\varepsilon \in \mathbb{R}$, the group of all such symmetries is uncountably infinite.

2.2.2 Lie symmetries of differential and difference equations

A transformation[6] of a differential or difference equation is a *symmetry* if every solution of the transformed equation is a solution of original equation and vice versa. Thus the set of all solutions is mapped invertibly to itself.

So far, we have considered only translations, reflections and rotations, all of which are rigid. However, many symmetries of differential and difference equations are not rigid. The freedom to deform the space of continuous variables smoothly turns out to be an essential ingredient in symmetry methods.

Example 2.4   Consider the effect of a scaling transformation,

$$\Gamma_\varepsilon : u \mapsto \hat{u} = e^\varepsilon u, \tag{2.8}$$

on a scalar linear homogeneous OΔE of order p (see §1.7 for the case $p = 2$). If $U_1(n), \ldots, U_p(n)$ are linearly independent solutions, the general solution is

$$u = \sum_{i=1}^{p} c_i U_i(n).$$

The scaling (2.8) maps this solution to

$$\hat{u} = \sum_{i=1}^{p} \hat{c}_i U_i(n), \qquad \text{where} \quad \hat{c}_i = e^\varepsilon c_i,$$

so the set of all solutions is mapped (invertibly) to itself. Thus Γ_ε is a symmetry of the OΔE for every $\varepsilon \in \mathbb{R}$. ▲

In the above example, $\hat{u}$ is a smooth function of u. Indeed, $\Gamma_\varepsilon : \mathbb{R} \to \mathbb{R}$ is a *diffeomorphism*, that is, a smooth invertible mapping whose inverse is also smooth. It is easy to check that the set of transformations $G = \{\Gamma_\varepsilon : \varepsilon \in \mathbb{R}\}$ is a group with the composition $\Gamma_\delta \Gamma_\varepsilon = \Gamma_{\delta+\varepsilon}$ for all $\delta, \varepsilon \in \mathbb{R}$; here Γ_0 is the identity map and $\Gamma_\varepsilon^{-1} = \Gamma_{-\varepsilon}$. Furthermore, $\hat{u}$ is an analytic function of the parameter ε.

This group has a very useful property: Γ_ε is a near-identity transformation for all sufficiently small ε. If these near-identity transformations are

---

[6] There are some technical restrictions on the class of allowable transformations; these will be discussed as they arise.

symmetries of a given p^{th}-order O∆E or ODE, a single solution will be mapped to a one-parameter family of nearby solutions whose arbitrary constants depend analytically on ε. This property can be used to solve first-order O∆Es (see §2.3).

For a nonlinear O∆E or ODE, a set of symmetries that depend continuously on a parameter may only be able to alter the arbitrary constants in the general solution locally (that is, by a sufficiently small amount). This is because some values of the arbitrary constants can produce blow-up. These values must be avoided; they are not part of the general solution.

Example 2.5 In Example 2.8, we shall show that the general solution of

$$u_1 = \frac{u}{1 + nu}, \qquad n \geq 1,$$

is

$$u = \frac{2}{n(n-1) - 2c_1}.$$

Such a solution will blow up whenever $c_1 = n_1(n_1 - 1)/2$, for $n_1 \in \mathbb{N}$. This restricts how far c_1 can be varied continuously. ▲

To obtain one-parameter families of symmetries that avoid blow-up, it is sufficient to look for near-identity transformations of the continuous variable(s). Given an equation (or system) whose continuous variables are $\mathbf{x} = (x^1, \ldots, x^N)$, a *point transformation* is a locally defined diffeomorphism $\Gamma : \mathbf{x} \mapsto \hat{\mathbf{x}}(\mathbf{x})$; the term 'point' is used because $\hat{\mathbf{x}}$ depends only on the point $\mathbf{x}$. A parametrized set of point transformations,

$$\Gamma_\varepsilon : \mathbf{x} \mapsto \hat{\mathbf{x}}(\mathbf{x}; \varepsilon), \qquad \varepsilon \in (\varepsilon_0, \varepsilon_1),$$

where $\varepsilon_0 < 0$ and $\varepsilon_1 > 0$, is a *one-parameter local Lie group* if the following conditions are satisfied:

1. Γ_0 is the identity map, so that $\hat{\mathbf{x}} = \mathbf{x}$ when $\varepsilon = 0$.
2. $\Gamma_\delta \Gamma_\varepsilon = \Gamma_{\delta+\varepsilon}$ for every δ, ε sufficiently close to zero.
3. Each $\hat{x}^\alpha$ can be represented as a Taylor series in ε (in a neighbourhood of $\varepsilon = 0$ that is determined by $\mathbf{x}$), and therefore

$$\hat{x}^\alpha(\mathbf{x}; \varepsilon) = x^\alpha + \varepsilon \xi^\alpha(\mathbf{x}) + O(\varepsilon^2), \qquad \alpha = 1, \ldots, N.$$

Conditions **1** and **2** ensure that $\Gamma_\varepsilon^{-1} = \Gamma_{-\varepsilon}$ when $|\varepsilon|$ is sufficiently small. Despite its name, a local Lie group may not be a group; it need only satisfy the group axioms for sufficiently small parameter values.

In general, a one-parameter local Lie group of symmetries of a given scalar O∆E will depend on n as well as on the continuous variable u, as the following example shows.

Example 2.6 The general solution of the linear OΔE

$$u_1 = \frac{n+1}{n} u, \qquad n \geq 1,$$

is $u = c_1 n$. Every transformation of the form

$$(\hat{n}, \hat{u}) = (n, u + \varepsilon n) \qquad (2.9)$$

is a symmetry that maps $u = c_1 n$ to $\hat{u} = (c_1 + \varepsilon)n$. For each $n \geq 1$, (2.9) defines a one-parameter local Lie group of translations; however, the amount by which u is translated for a given ε varies with n. ▲

For brevity, we refer to symmetries that belong to a one-parameter local Lie group (for each n, in the case of OΔEs) as *Lie symmetries*. Similarly, *Lie point transformations* are point transformations that belong to a one-parameter local Lie group.

2.2.3 Not all symmetries change solutions

Most methods for solving a given ODE use a single idea, which is to transform the problem into something simpler (typically, by changing variables). Roughly, this amounts to finding some solutions, then using Lie symmetries to map these to the remaining solutions. For this approach to work, the Lie symmetries must map almost all solutions to different solutions. However, some symmetries do not change any solutions; these are called *trivial* symmetries.

Unlike ODEs, OΔEs have no trivial Lie symmetries. To see why, it is helpful to compare the solutions of the simplest ODE,

$$\frac{dy}{dx} = 0, \qquad (2.10)$$

with those of the simplest OΔE,

$$u_1 - u = 0. \qquad (2.11)$$

The solutions of the ODE (2.10) are the lines $y = c_1$, as shown in Figure 2.4(a), so the symmetries of (2.10) map each such line to another one. Among the Lie symmetries are vertical and horizontal translations,

$$(\hat{x}, \hat{y}) = (x, y + \varepsilon_1), \qquad \varepsilon_1 \in \mathbb{R}; \qquad (2.12)$$

$$(\hat{x}, \hat{y}) = (x + \varepsilon_2, y), \qquad \varepsilon_2 \in \mathbb{R}. \qquad (2.13)$$

The vertical translations (2.12) map the solution $y = c_1$ to $y = c_1 + \varepsilon_1$; by varying ε_1, one can obtain every solution of the ODE from a single solution. By contrast, the horizontal translations (2.13) are trivial. The fact that (2.13)

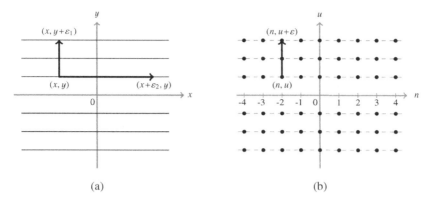

Figure 2.4 (a) Some solutions of $dy/dx = 0$, with arrows showing the effect of the symmetries (2.12) and (2.13) on an arbitrary point (x, y). (b) The corresponding figure for the OΔE $u_1 - u = 0$ and the symmetries (2.14) – the dashed lines show which points (shown as blobs) belong to which solution.

moves points along each solution curve is immaterial; the curves themselves are unaffected.

For the OΔE (2.11), some solutions, $u = c_1$, are shown in Figure 2.4(b). Once again, the Lie symmetries include vertical translations,

$$(\hat{n}, \hat{u}) = (n, u + \varepsilon), \qquad \varepsilon \in \mathbb{R}. \tag{2.14}$$

Just as for the ODE (2.10), every such symmetry with $\varepsilon \neq 0$ maps each solution, $u = c_1$, to a different solution, $u = c_1 + \varepsilon$. But n is a discrete variable that cannot be changed by an arbitrarily small amount, so every one-parameter local Lie group of symmetries must leave n unchanged. In other words, $\hat{n} = n$ for all Lie symmetries of (2.11). The same applies to all OΔEs; consequently, OΔEs have no trivial Lie symmetries[7] that are point transformations of (n, u).

2.2.4 Characteristics and canonical coordinates

For the remainder of this chapter, we restrict attention to Lie symmetries for which $\hat{u}$ depends on n and u only. These are called *Lie point symmetries*; they are of the form

$$\hat{n} = n, \qquad \hat{u} = u + \varepsilon Q(n, u) + O(\varepsilon^2). \tag{2.15}$$

---

[7] Equation (2.11) does have some trivial symmetries that are point transformations, including $(\hat{n}, \hat{u}) = (n + 1, u)$, but these are all discrete symmetries, not Lie symmetries (see Chapter 4).

To see how such symmetries transform the shifted variables u_k, simply replace the free variable n in (2.15) by $n + k$:

$$\hat{u}_k = u_k + \varepsilon Q(n + k, u_k) + O(\varepsilon^2). \tag{2.16}$$

This is the *prolongation formula* for Lie point symmetries.

The function $Q(n, u)$ is called the *characteristic* of the local Lie group with respect to the coordinates (n, u). For instance, the characteristic that corresponds to the vertical translations (2.14) is

$$Q(n, u) = 1. \tag{2.17}$$

Now consider the effect of changing coordinates from (n, u) to (n, v), where $v'(n, u) \neq 0$; here $v' = \partial v / \partial u$. (More generally, given any function F that depends on n and a single continuous variable a, we use F' to denote the derivative with respect to a.) As (2.15) is a symmetry for each ε sufficiently close to zero, we can apply Taylor's Theorem to obtain

$$\hat{v} \equiv v(n, \hat{u}) = v(n, u + \varepsilon Q(n, u) + O(\varepsilon^2)) = v + \varepsilon v'(n, u) Q(n, u) + O(\varepsilon^2).$$

Therefore the characteristic with respect to (n, v) is $\tilde{Q}(n, v)$, where

$$\tilde{Q}(n, v(n, u)) = v'(n, u) Q(n, u). \tag{2.18}$$

At most points (n, u) where $v'(n, u) \neq 1$, the values of $\tilde{Q}$ and Q will differ; the only exceptions are points where $Q(n, u) = 0$. Where $Q(n, u) \neq 0$, it is particularly useful to introduce a *canonical coordinate*, s, such that the symmetries amount to translations in s:

$$(\hat{n}, \hat{s}) = (n, s + \varepsilon), \qquad \varepsilon \in \mathbb{R}. \tag{2.19}$$

The characteristic with respect to (n, s) is $\tilde{Q}(n, s) = 1$, and so, by (2.18),

$$s(n, u) = \int \frac{du}{Q(n, u)}. \tag{2.20}$$

For each n, the possible values of u lie on the real line, which (typically) splits into intervals when we omit every value u for which $Q(n, u) = 0$. Equation (2.20) defines a canonical coordinate s (locally) on each interval, but it may happen that different coordinates are appropriate for different intervals. For instance, if $Q(n, u) = u^2 - 1$, the appropriate real-valued canonical coordinate depends on u, as follows:

$$s(n, u) = \int \frac{du}{u^2 - 1} = \begin{cases} \frac{1}{2} \ln \{(u - 1)/(u + 1)\}, & |u| > 1; \\ \frac{1}{2} \ln \{(1 - u)/(1 + u)\}, & |u| < 1. \end{cases}$$

In this example, $s(n, 1/u) = s(n, u)$ for each nonzero u, so the map from u to s

is not injective; it cannot be inverted unless one specifies in advance whether $|u|$ is greater or less than 1.

One of the main uses of a canonical coordinate is to simplify or even solve a given OΔE. The idea is to rewrite the OΔE as a simpler OΔE for s; if the simpler OΔE can be solved, all that remains is to write the solution in terms of the original variables. To use this approach, therefore, one must be able to invert the map from u to s (at least, for all points (n, u) that occur in any solution of the original OΔE and satisfy $Q(n, u) \neq 0$). Any canonical coordinate s that meets this requirement will be called *compatible* with the OΔE.

For any compatible canonical coordinate, we can replace n by $n + k$ to obtain

$$s_k = s(n + k, u_k) = S_n^k s, \qquad k \in \mathbb{Z}.$$

By the prolongation formula, the Lie symmetry $\hat{s} = s + \varepsilon$ prolongs to

$$\hat{s}_k = s_k + \varepsilon.$$

2.3 Lie symmetries solve first-order OΔEs

At this stage, we have enough tools to solve a given first-order OΔE,

$$u_1 = \omega(n, u), \tag{2.21}$$

by using a one-parameter local Lie group of symmetries. For the set of solutions of (2.21) to be mapped to itself, the following *symmetry condition* must be satisfied:

$$\hat{u}_1 = \omega(\hat{n}, \hat{u}) \quad \text{when} \quad u_1 = \omega(n, u). \tag{2.22}$$

Example 2.7   In Chapter 1, we saw that the simplest class of OΔEs,

$$u_1 - u = b(n), \tag{2.23}$$

can be solved by summation alone. Each OΔE in this class has the following one-parameter Lie group of symmetries, prolonged according to (2.16):

$$(\hat{n}, \hat{u}, \hat{u}_1) = (n, u + \varepsilon, u_1 + \varepsilon). \tag{2.24}$$

The symmetry condition is satisfied for each $\varepsilon \in \mathbb{R}$ because, on solutions of (2.23),

$$\hat{u}_1 - \hat{u} = (u_1 + \varepsilon) - (u + \varepsilon) = u_1 - u = b(n) = b(\hat{n}).$$

Furthermore, it is an easy exercise to show that no other first-order OΔE (2.21) has these Lie symmetries. ▲

Although Example 2.7 seems particularly simple, it is the key to finding the general solution of the OΔE (2.21). Given a nonzero characteristic, $Q(n, u)$, suppose that we can write the OΔE in terms of a single compatible canonical coordinate s (except where $Q(n, u)$ or $Q(n + 1, u_1)$ is zero). As the Lie symmetries amount to translations in the canonical coordinate, this change of variables maps the OΔE to

$$s_1 - s = b(n); \tag{2.25}$$

here $b(n)$ is found by evaluating the left-hand side of (2.25) explicitly. The general solution of (2.25) is

$$s = \sigma_k\{b(k); n_0, n\} + c_1,$$

where n_0 is any convenient integer. All that remains is to rewrite this solution in terms of n and u, which can be done (in principle) because the map from u to s is injective.

Example 2.8 The OΔE

$$u_1 = \frac{u}{1 + nu}, \qquad n \geq 1,$$

has the one-parameter Lie group of symmetries

$$(\hat{n}, \hat{u}) = \left(n, \frac{u}{1 - \varepsilon u}\right),$$

whose characteristic is $Q(n, u) = u^2$ (check this). For this characteristic,

$$s(n, u) = \int u^{-2} du = -u^{-1}$$

is compatible with the OΔE, which amounts to

$$s_1 - s = -(u_1)^{-1} + u^{-1} = -n,$$

and hence

$$s = c_1 - \sum_{k=1}^{n-1} k = c_1 - n(n - 1)/2.$$

Consequently, the general solution of the original OΔE is

$$u = \frac{2}{n(n - 1) - 2c_1}.$$

The coordinate s cannot be used at $u = 0$, as $Q(n, 0) = 0$. Inspecting the OΔE, we see that $u = 0$ is a solution, although it is not part of the general solution. ▲

Example 2.9 For the last example, we could easily have spotted that the OΔE is linearizable, without having to know a characteristic. As a less obvious example, consider the OΔE

$$u_1 = \frac{nu + 1}{u + n}, \qquad n \geq 2, \qquad u(2) \geq -1.$$

Again, there is a characteristic that is quadratic in u, namely $Q(n, u) = u^2 - 1$. For this characteristic, no canonical coordinate exists at $u = \pm 1$. Clearly, if $u(2) = \pm 1$ then $u(n) = u(2)$ for all $n \geq 2$. Moreover, if u belongs to either $(1, \infty)$ or $(-1, 1)$ then u_1 belongs to the same interval. Therefore the problem of solving the OΔE (away from $u = \pm 1$) splits into two separate parts.

For now, restrict attention to the case $u(2) > 1$; in this case, $u > 1$ for all $n \geq 2$, and for $s(n, u) = \frac{1}{2} \ln \{(u - 1)/(u + 1)\}$, the transformation from u to s is injective. The transformed OΔE is

$$s_1 - s = \frac{1}{2} \ln \left(\frac{n - 1}{n + 1} \right) = \frac{1}{2} \ln (n(n - 1)) - \frac{1}{2} \ln (n(n + 1)),$$

and so

$$s = \frac{1}{2} \ln \left(\frac{2(u(2) - 1)}{(u(2) + 1)n(n - 1)} \right).$$

Thus the solution of the original OΔE when $u(2) > 1$ is

$$u = \frac{(u(2) + 1)n(n - 1) + 2(u(2) - 1)}{(u(2) + 1)n(n - 1) - 2(u(2) - 1)}.$$

In fact, this solution is valid for all $u(2) \geq -1$. When $|u(2)| < 1$, we use a canonical coordinate that suits $|u| < 1$, namely $\tilde{s} = \frac{1}{2} \ln \{(1 - u)/(1 + u)\}$; the details of the calculation are left to the reader. The general solution happens to include the solutions on which $Q(n, u) = 0$. ▲

In the example above we introduced a modification to the basic method: if one cannot find a compatible canonical coordinate, split the OΔE into separate pieces with a compatible canonical coordinate on each piece. But what happens if one needs to use two different canonical coordinates for u and u_1? For instance, suppose that s and $\tilde{s}$ are needed when $u < 0$ and $u > 0$, respectively. If $u(n) < 0$ for $n < m_0$ and $u(n) > 0$ otherwise, one can again split the problem of solving the OΔE into two parts, dealing with the awkward point $n = m_0 - 1$ by hand. But this becomes a laborious process when there are multiple crossings between regions where different canonical coordinates apply.

Such problems can often be resolved by allowing s to take complex values. If this creates the possibility of s having multiple values, use suitable branch cuts (for instance, choose the principal value).

Example 2.10   Consider the OΔE

$$u_1 = \frac{u - n}{nu - 1}, \qquad n \geq 2,$$

which has a characteristic $Q(n, u) = (-1)^n(u^2 - 1)$. It is easy to show that $|u_1|$ is greater (less) than 1 whenever $|u|$ is less (greater) than 1 and $u \neq 1/n$. Although no real-valued compatible s exists, the problem of writing the OΔE in terms of a compatible canonical coordinate is overcome when we use

$$s(n, u) = \frac{(-1)^n}{2} \mathrm{Log}\left(\frac{u - 1}{u + 1}\right) = \begin{cases} \dfrac{(-1)^n}{2} \ln\left(\dfrac{u - 1}{u + 1}\right), & \dfrac{u - 1}{u + 1} > 0; \\[3mm] \dfrac{(-1)^n}{2}\left(\ln\left(\dfrac{1 - u}{1 + u}\right) + i\pi\right), & \dfrac{u - 1}{u + 1} < 0. \end{cases}$$

Here Log is the principal value of the complex logarithm:

$$\mathrm{Log}(z) = \ln(|z|) + i\mathrm{Arg}(z), \qquad \mathrm{Arg}(z) \in (-\pi, \pi].$$

Whether $(u - 1)/(u + 1)$ is positive or negative, the OΔE amounts to

$$s_1 - s = \tfrac{1}{2}(-1)^{n+1}\{\ln((n - 1)/(n + 1)) + i\pi\},$$

whose general solution is

$$s = \tfrac{1}{2}(-1)^n\{\ln((n - 1)/n) + i\pi/2\} + c_1,$$

where c_1 is a complex-valued constant that can be written in terms of $u(2)$. A routine calculation yields the general solution of the original OΔE:

$$u = \begin{cases} \dfrac{(u(2) + 1)n + 2(u(2) - 1)(n - 1)}{(u(2) + 1)n - 2(u(2) - 1)(n - 1)}, & n \text{ even}; \\[3mm] \dfrac{2(1 - u(2))n + (u(2) + 1)(n - 1)}{2(1 - u(2))n - (u(2) + 1)(n - 1)}, & n \text{ odd}. \end{cases} \qquad \blacktriangle$$

2.4 How to find Lie symmetries of a given OΔE

2.4.1 First-order OΔEs

Now that we know how to use a characteristic to solve a given OΔE, it is natural to ask: how can we find such a characteristic? Generally speaking, the symmetry condition is hard to solve by brute force. For Lie symmetries, however, the symmetry condition can be expanded in powers of ε (for ε sufficiently close to zero) as follows:

$$\{u_1 + \varepsilon Q(n+1, u_1) + O(\varepsilon^2)\}\Big|_{u_1 = \omega(n,u)} = \omega(n, u + \varepsilon Q(n, u) + O(\varepsilon^2))$$

$$= \omega(n, u) + \varepsilon \omega'(n, u) Q(n, u) + O(\varepsilon^2).$$

The $O(1)$ terms cancel, so the leading-order constraint appears at $O(\varepsilon)$:

$$Q(n+1, \omega(n, u)) = \omega'(n, u) Q(n, u). \tag{2.26}$$

This constraint on $Q(n, u)$ is called the *linearized symmetry condition* (LSC) for the OΔE (2.21). Unlike the symmetry condition, the LSC is linear whether or not the OΔE is linear. Thus, for nonlinear OΔEs, it is generally easier to find Lie symmetries than discrete symmetries.

The LSC (2.26) is a linear functional equation that appears to be difficult to solve completely. Appearances can be misleading, however.

Example 2.11 The LSC for $u_1 = u$ is $Q(n+1, u) = Q(n, u)$, whose general solution is $Q(n, u) = F(u)$, where F is an arbitrary function. The OΔE

$$u_1 - u = b(n) \tag{2.27}$$

can be written as $v_1 = v$, where $v = u - \sigma_k\{b(k); n_0, n\}$ for any convenient n_0. So, by the change-of-variables formula (2.18), the general solution of the LSC for (2.27) is

$$Q(n, u) = F(u - \sigma_k\{b(k); n_0, n\}). \qquad \blacktriangle$$

Example 2.12 The linear homogeneous OΔE $u_1 = -a(n) u$ has the LSC

$$Q(n+1, -a(n) u) = -a(n) Q(n, u).$$

Some solutions are obvious, such as

$$Q(n, u) = c_1 u.$$

Just as in the last example, the key to finding the general solution is to write the OΔE in the form $v_1 = v$; in this case,

$$v = \frac{u}{\pi_l\{-a(l); n_0, n\}}.$$

We already know that $v_1 = v$ has Lie symmetries whose characteristic is an arbitrary function of v. The change-of-variables formula for characteristics gives the general solution of the LSC in the original variables:

$$Q(n, u) = \pi_l\{-a(l); n_0, n\} F\left(\frac{u}{\pi_l\{-a(l); n_0, n\}}\right). \qquad \blacktriangle$$

In the same way, one can always find the general solution of the LSC if one knows the general solution of the OΔE (2.21) – see Exercise 2.10. This is not particularly useful, because most OΔEs do not belong to a class that we know how to solve. However, there is no real need to find the general solution of the LSC, as a single nonzero solution of the LSC is sufficient to determine the general solution of the OΔE. A practical approach is to use an ansatz (trial solution). Many physically important Lie point symmetries have characteristics of the form[8]

$$Q(n, u) = \alpha(n)\, u^2 + \beta(n)\, u + \gamma(n). \tag{2.28}$$

By substituting (2.28) into the LSC and comparing powers of u, one obtains a linear system of OΔEs for the coefficients $\alpha(n)$, $\beta(n)$ and $\gamma(n)$. The aim is to seek at least one nonzero solution of this system; it may be necessary to introduce ansätze for $\alpha(n)$, $\beta(n)$ and $\gamma(n)$ if the system cannot be solved in full generality.

Example 2.13   Use the ansatz (2.28) to seek Lie point symmetries of

$$u_1 = \frac{u}{1 + nu}, \qquad n \geq 1.$$

Solution: The LSC for the above OΔE is

$$Q\left(n + 1, \frac{u}{1 + nu}\right) = \frac{1}{(1 + nu)^2}\, Q(n, u).$$

With the ansatz (2.28), the LSC reduces to the OΔE

$$u^2\alpha_1 + u(1 + nu)\beta_1 + (1 + nu)^2\gamma_1 = u^2\alpha + u\beta + \gamma,$$

which we split into a system of OΔEs by comparing powers of u:

$$u^2 \text{ terms} : \qquad\qquad \alpha_1 + n\beta_1 + n^2\gamma_1 = \alpha,$$
$$u \text{ terms} : \qquad\qquad \beta_1 + 2n\gamma_1 = \beta,$$
$$\text{other terms} : \qquad\qquad \gamma_1 = \gamma.$$

Systems in this 'triangular' form are solved by starting at the bottom and working up. In this case, each of the scalar equations is easy to solve, so there is no need to introduce additional ansätze for α, β or γ. We obtain

$$\gamma = c_1,$$
$$\beta = c_2 - c_1 n(n - 1),$$
$$\alpha = c_3 - c_2 n(n - 1)/2 + c_1 n^2(n - 1)^2/4.$$

---

[8]   Indeed, for nearly every OΔE of order $p \geq 2$, there exists a local change of coordinates that puts all characteristics into this form – see Chapter 3, in Notes and further reading.

Therefore every characteristic that is quadratic in u is of the form

$$Q(n,u) = c_1\left(1 - \frac{n(n-1)}{2}u\right)^2 + c_2 u\left(1 - \frac{n(n-1)}{2}u\right) + c_3 u^2. \qquad \blacktriangle$$

Example 2.14 Let us revisit the OΔE from Example 2.10,

$$u_1 = \frac{u-n}{nu-1}, \qquad n \geq 2,$$

to see whether the ansatz (2.28) will yield any new characteristics. The LSC,

$$Q\left(n+1, \frac{u-n}{nu-1}\right) = \frac{n^2-1}{(nu-1)^2} Q(n,u),$$

reduces to the OΔE

$$(u-n)^2\alpha_1 + (u-n)(nu-1)\beta_1 + (nu-1)^2\gamma_1 = (n^2-1)\{u^2\alpha + u\beta + \gamma\}.$$

This splits as follows:

$$\begin{aligned}
u^2 \text{ terms}: && \alpha_1 + n\beta_1 + n^2\gamma_1 &= (n^2-1)\alpha, \\
u \text{ terms}: && -2n\alpha_1 - (n^2+1)\beta_1 - 2n\gamma_1 &= (n^2-1)\beta, \\
\text{other terms}: && n^2\alpha_1 + n\beta_1 + \gamma_1 &= (n^2-1)\gamma.
\end{aligned}$$

Most highly coupled (that is, non-triangular) linear systems with non-constant coefficients are hard to solve, so it is tempting to resort to ansätze for α, β and γ. However, it is always worth looking first to see whether the system can be simplified by forming linear combinations of the equations, an approach that yields dividends in this case. A little manipulation produces the following system:

$$\alpha_1 - \gamma_1 = -(\alpha - \gamma), \qquad (2.29)$$

$$(n+1)(\alpha_1 + \beta_1 + \gamma_1) = (n-1)(\alpha - \beta + \gamma), \qquad (2.30)$$

$$(n-1)(\alpha_1 - \beta_1 + \gamma_1) = (n+1)(\alpha + \beta + \gamma). \qquad (2.31)$$

The general solution of (2.29) is

$$\alpha - \gamma = c_1(-1)^n.$$

Equations (2.30) and (2.31) together yield the second-order scalar OΔE

$$\frac{n+2}{n+1}(\alpha_2 + \beta_2 + \gamma_2) = \frac{n}{n-1}(\alpha + \beta + \gamma),$$

whose general solution is

$$\alpha + \beta + \gamma = \frac{n-1}{n}(c_2 + c_3(-1)^n).$$

From this, (2.30) supplies the final equation for α, β and γ, namely

$$\alpha - \beta + \gamma = \frac{n}{n-1} \left(c_2 - c_3(-1)^n \right).$$

Solving for α, β and γ (and rescaling the constants, for convenience), we obtain

$$Q(n,u) = \tilde{c}_1(-1)^n \left(1 - u^2 \right) + \tilde{c}_2 \left\{ \frac{n-1}{n} (1+u)^2 + \frac{n}{n-1} (1-u)^2 \right\}$$
$$+ \tilde{c}_3(-1)^n \left\{ \frac{n-1}{n} (1+u)^2 - \frac{n}{n-1} (1-u)^2 \right\}. \qquad \blacktriangle$$

2.4.2 Second- and higher-order OΔEs

The linearized symmetry condition for OΔEs of order $p > 1$ is obtained by the same approach as we used for first-order OΔEs. For definiteness, we will consider only forward OΔEs,

$$u_p = \omega(n, u, \dots, u_{p-1}), \qquad n \in D; \tag{2.32}$$

we assume that $\omega_{,1} \neq 0$ for each $n \in D$, so that D is a regular domain. The symmetry condition is

$$\hat{u}_p = \omega(\hat{n}, \hat{u}, \dots, \hat{u}_{p-1}) \qquad \text{when (2.32) holds.} \tag{2.33}$$

As before, substitute

$$\hat{n} = n, \qquad \hat{u} = u + \varepsilon Q(n, u) + O(\varepsilon^2), \qquad \hat{u}_k = u_k + \varepsilon Q(n + k, u_k) + O(\varepsilon^2),$$

into (2.33) to obtain the LSC:

$$Q(n + p, u_p) = \sum_{k=0}^{p-1} \omega_{,k+1} Q(n + k, u_k) \qquad \text{when (2.32) holds.} \tag{2.34}$$

Note: We are assuming that if ε is sufficiently small, $\hat{u}_k - u_k = O(\varepsilon)$ whenever $\hat{u} - u = O(\varepsilon)$.

Just as for first-order OΔEs, the LSC is a linear functional equation for the characteristics Q. However, for most nonlinear OΔEs of order $p > 1$, one can find all Lie point symmetries from (2.34), without having to use ansätze. To see how this works, let us examine the LSC for a given second-order OΔE,

$$u_2 = \omega(n, u, u_1), \tag{2.35}$$

which is

$$Q(n + 2, \omega) - \omega_{,2} Q(n + 1, u_1) - \omega_{,1} Q(n, u) = 0. \tag{2.36}$$

Although functional equations are generally hard to solve, Lie symmetries are diffeomorphisms. Consequently, Q is a smooth function of its continuous argument and so the LSC can be solved by the method of *differential elim-ination*. Before discussing this method in general, we begin with a simple example.

Example 2.15 Find the characteristics of Lie point symmetries for

$$u_2 = (u_1)^2/u.$$

Here the LSC (2.36) is

$$Q\left(n+2, (u_1)^2/u\right) - 2(u_1/u)\, Q(n+1, u_1) + (u_1/u)^2\, Q(n, u) = 0. \quad (2.37)$$

In each term, Q depends on only one continuous variable (and one discrete variable). Therefore it is possible to differentiate the LSC in order to eliminate that term. To begin with, note that

$$\left(-\frac{u^2}{(u_1)^2}\frac{\partial}{\partial u} - \frac{u}{2u_1}\frac{\partial}{\partial u_1}\right)Q\left(n+2, (u_1)^2/u\right) = 0.$$

By applying the differential operator in parentheses to the LSC, we eliminate the first term in (2.37) and obtain the functional-differential equation

$$Q'(n+1, u_1) - \frac{1}{u_1}\, Q(n+1, u_1) - Q'(n, u) + \frac{1}{u}\, Q(n, u) = 0. \quad (2.38)$$

This has more terms than the LSC does, but Q and Q' only take two pairs of arguments, instead of three. Now we can continue the process, by differentiating (2.38) with respect to u, in order to eliminate $Q'(n+1, u_1)$. This gives

$$\left[-Q'(n, u) + \frac{1}{u}\, Q(n, u)\right]' = 0, \quad (2.39)$$

which is a parametrized ODE. At this stage Q and its derivatives depend upon only one pair of arguments, so we can start to solve the system. Integrate (2.39) *once*, to obtain

$$Q'(n, u) - \frac{1}{u}\, Q(n, u) = \alpha(n), \quad (2.40)$$

where $\alpha(n)$ is to be determined. Now substitute (2.40) into the intermediate equation (2.38), which gives the OΔE

$$\alpha_1 - \alpha = 0.$$

Therefore $\alpha = c_1$, so the general solution of (2.40) is

$$Q(n, u) = c_1 u \ln |u| + \beta(n)\, u. \quad (2.41)$$

Finally, substitute (2.41) into the original LSC (2.37) to obtain (after cancellation)

$$\beta_2 - 2\beta_1 + \beta = 0.$$

The general solution of this linear OΔE is

$$\beta(n) = c_2 n + c_3,$$

and hence the general solution of the LSC is

$$Q(n, u) = c_1 u \ln |u| + c_2 n u + c_3 u. \qquad \blacktriangle$$

The above example illustrates the steps that are used in the method of differential elimination. Beginning with the LSC, one constructs a sequence of intermediate equations by using first-order partial differential operators to eliminate terms. Each intermediate equation is satisfied by every sufficiently differentiable[9] solution of the LSC. Eventually, this process yields a differential equation for $Q(n, u)$, which is integrated one order at a time. After each integration, the result is substituted into the previous equation in the sequence; typically, this generates a linear difference equation, which should be solved if possible. (It might be necessary to wait until further information is available.) In the end, one arrives back at the LSC, which provides the final constraints.

From here on, we consider only OΔEs that satisfy $\omega_{,2} \neq 0$, together with the regularity condition $\omega_{,1} \neq 0$. (Exercise 1.16 gives a way of dealing with second-order OΔEs for which $\omega_{,2} = 0$; see also §3.6.)

The first term of the LSC (2.36) is eliminated by applying the differential operator $(1/\omega_{,1}) \partial/\partial u - (1/\omega_{,2}) \partial/\partial u_1$, which yields

$$Q'(n + 1, u_1) + \eta_{,2} Q(n + 1, u_1) - Q'(n, u) + \eta_{,1} Q(n, u) = 0, \qquad (2.42)$$

where $\eta(n, u, u_1) = \ln |\omega_{,2}/\omega_{,1}|$. Next, one can eliminate $Q'(n + 1, u_1)$ by differentiating (2.42) with respect to u:

$$\eta_{,12} Q(n + 1, u_1) - Q''(n, u) + \eta_{,1} Q'(n, u) + \eta_{,11} Q(n, u) = 0. \qquad (2.43)$$

At this stage, there are two possibilities.

First, if $\eta_{,12} = 0$, (2.43) can be integrated once to obtain

$$Q'(n, u) - \eta_{,1} Q(n, u) = \alpha(n). \qquad (2.44)$$

Now substitute (2.44) into (2.42), to eliminate $Q'(n, u)$ and $Q'(n + 1, u_1)$:

$$(S_n \eta_{,1} + \eta_{,2}) Q(n + 1, u_1) = \alpha(n) - \alpha(n + 1). \qquad (2.45)$$

If $\eta_{,2} = -S_n \eta_{,1}$ then (2.45) implies that $\alpha(n) = c_1$ and (2.44) can be integrated to obtain $Q(n, u)$. Otherwise (2.45) yields

$$Q(n, u) = \frac{\alpha(n - 1) - \alpha(n)}{\eta_{,1} + S_n^{-1} \eta_{,2}}, \qquad (2.46)$$

---

[9] We do not consider pathological cases, as the purpose of this book is to present the basics of the subject, without dwelling on every technicality.

which can be substituted into (2.44), giving (at most) a first-order linear OΔE for $\alpha(n)$. In either case, the function $Q(n, u)$ that results from (2.42) must be substituted into the LSC (2.36), and one must solve any extra constraints that this produces.

The second possibility is that $\eta_{,12} \neq 0$, in which case (2.43) should be divided by $\eta_{,12}$ and then differentiated once more with respect to u. The coefficients of the resulting OΔE may depend upon u_1; if this happens, the OΔE must be split into a system of OΔEs whose coefficients are functions of n and u only. Then the solution process continues as before.

Example 2.16 Find all characteristics of the Lie point symmetries for the OΔE

$$u_2 = (u_1)^2 + u. \tag{2.47}$$

Solution: Here $\omega_{,1} = 1$ and $\omega_{,2} = 2u_1$; consequently, $\eta = \ln |2u_1|$ and $\eta_{,12} = 0$. In this case, (2.46) applies and we obtain

$$Q(n, u) = (\alpha(n - 1) - \alpha(n))u.$$

Therefore (2.44) amounts to the OΔE

$$\alpha_{-1} - \alpha = \alpha,$$

whose solution is $\alpha = c_1 2^{-n}$. Finally, substitute $Q(n, u) = c_1 2^{-n} u$ into the LSC,

$$Q\left(n + 2, (u_1)^2 + u\right) - 2u_1 Q(n + 1, u_1) - Q(n, u) = 0,$$

which is the only remaining constraint. This gives

$$c_1 2^{-(n+2)} \left((u_1)^2 + u\right) - c_1 2^{-n}(u_1)^2 - c_1 2^{-n} u = 0,$$

and so $c_1 = 0$. In other words, the OΔE (2.47) has no Lie point symmetries. ▲

Example 2.17 Find all characteristics of Lie point symmetries for the OΔE

$$u_2 = \frac{1}{u_1 + u} - u_1 - 2(-1)^n, \qquad n \geq 0. \tag{2.48}$$

Solution: (This is going to get a little messy!) The LSC is

$$Q\left(n + 2, \frac{1}{u_1 + u} - u_1 - 2(-1)^n\right) + \left(1 + \frac{1}{(u_1 + u)^2}\right) Q(n+1, u_1) + \frac{Q(n, u)}{(u_1 + u)^2} = 0,$$

and $\eta = \ln\left((u_1 + u)^2 + 1\right)$. As $\eta_{,12} \neq 0$, we divide (2.43) by $\eta_{,12}$, obtaining

$$Q(n + 1, u_1) + \frac{\left((u_1 + u)^2 + 1\right)^2}{2\left((u_1 + u)^2 - 1\right)} Q''(n, u)$$
$$- \frac{(u_1 + u)\left((u_1 + u)^2 + 1\right)}{(u_1 + u)^2 - 1} Q'(n, u) + Q(n, u) = 0. \qquad (2.49)$$

We now eliminate $Q(n + 1, u_1)$ by differentiating the functional-differential equation (2.49) with respect to u; after simplification, this yields

$$\left((u_1 + u)^4 - 1\right) Q'''(n, u) - 4(u_1 + u)Q''(n, u) + 4Q'(n, u) = 0. \qquad (2.50)$$

The coefficients of (2.50) depend on u_1, so (2.50) splits into a set of ODEs, each of which is multiplied by a particular power of $(u_1 + u)$:

$$Q'''(n, u) = 0, \qquad Q''(n, u) = 0, \qquad Q'(n, u) = 0.$$

Therefore $Q(n, u) = \alpha(n)$. Substituting this result into (2.49) gives

$$\alpha_1 + \alpha = 0,$$

whose general solution is

$$Q(n, u) = c_1(-1)^n. \qquad (2.51)$$

The LSC is satisfied by this characteristic for every $c_1 \in \mathbb{R}$, so (2.51) is the general solution of the LSC. ▲

The same approach can be applied to finding all characteristics $Q(n, u)$ for OΔEs of order $p \geq 3$ that are not of the form $u_p = \omega(n, u)$ (this exclusion is explained in §3.6). With increasing order, the calculations grow rapidly in complexity, so it is essential to use computer algebra, but the process is simple.

(i) Write out the LSC for the OΔE.
(ii) Differentiate (and, if necessary, rearrange) the LSC repeatedly, so that at least one term involving a shift of $Q, Q', \ldots$ is eliminated. Continue doing this until an ODE is obtained.
(iii) Split the ODE into a system of ODEs whose coefficients involve only the arguments of the unknown function Q.
(iv) Simplify the system of ODEs (if possible).
(v) Integrate the simplified ODEs, one step at a time, and substitute the results successively into the hierarchy of functional-differential equations that were constructed in stage (ii). If possible, solve the linear OΔEs that arise.

(vi) Finally, substitute Q into the LSC, collect and simplify any remaining OΔEs for coefficients of terms in Q, and solve these.

In many instances, it is possible to find the general solution of the system of ODEs from stage (iv). Then one could proceed directly to stage (vi). In general, however, we cannot expect to be able to do this. Nevertheless, it is always possible to integrate one step at a time, because we know the differential operators that were applied at each step in stage (ii). The only part of the process that cannot be guaranteed is the solution of OΔEs for coefficients of terms in Q. These OΔEs are necessarily linear, because the LSC is linear in Q and its shifts. However, they may not be easy to solve.

Example 2.18   Find all characteristics of Lie point symmetries of

$$u_3 = f(n)\, u_1 + g(n)\, u,$$

where $f(n), g(n)$ are nonzero for all $n \in D$.

Solution: Here the LSC is

$$Q(n + 3, f(n)\, u_1 + g(n)\, u) - f(n)\, Q(n + 1, u_1) - g(n)\, Q(n, u) = 0.$$

Apply the differential operator $(1/g(n))\, \partial/\partial u - (1/f(n))\, \partial/\partial u_1$ to obtain

$$Q'(n + 1, u_1) - Q'(n, u) = 0.$$

Now apply $\partial/\partial u$, which yields the ODE

$$Q''(n, u) = 0.$$

Hence $Q'(n, u) = \alpha(n)$, where

$$\alpha_1 - \alpha = 0.$$

Therefore $Q(n, u) = c_1 u + \beta(n)$. The LSC gives the final constraint

$$\beta_3 - f(n)\beta_1 - g(n)\beta = 0.$$

In other words, $u = \beta(n)$ is the general solution of the original linear OΔE. So $Q(n, u)$ can be found explicitly only if the original OΔE can be solved. Nevertheless, we know that $\beta(n)$ involves three arbitrary constants, so we can conclude that there are four linearly independent characteristics $Q(n, u)$. More generally, any linear homogeneous OΔE of order p has characteristics of the form

$$Q(n, u) = c_1 u + \beta(n),$$

where $u = \beta(n)$ is the general solution of the OΔE. However, these are not necessarily the only characteristics of Lie point symmetries (see §3.6).   ▲

Note: In each of the examples so far, we have eliminated ω, then u_1, to leave an ODE for $Q(n, u)$. In general, this is a good approach, but it is not always the simplest way to derive an ODE for the characteristic. For some OΔEs, the calculations are simpler if one eliminates $Q(n, u)$ in order to find an ODE for $Q(n + i, u_i)$, for a particular $i \geq 1$.

2.5 Reduction of order for nonlinear OΔEs

To reduce the order of a given p^{th}-order linear ODE or OΔE, one needs to know a nonzero solution of the associated homogeneous equation. Sophus Lie extended this approach to nonlinear ODEs by exploiting one-parameter Lie groups of point symmetries. It turns out that Lie's method works equally well for nonlinear OΔEs, provided that one can find a compatible canonical coordinate. We start by adapting Lie's method to a given second-order OΔE,

$$u_2 = \omega(n, u, u_1). \tag{2.52}$$

Suppose that we have been able to find a characteristic $Q(n, u)$ for a given OΔE (2.52), and that s is a compatible canonical coordinate. Let

$$r = s_1 - s = \int \frac{du_1}{Q(n + 1, u_1)} - \int \frac{du}{Q(n, u)}. \tag{2.53}$$

Applying the shift operator to (2.53) gives

$$r_1 = \int \frac{du_2}{Q(n + 2, u_2)} - \int \frac{du_1}{Q(n + 1, u_1)}. \tag{2.54}$$

On solutions of the OΔE (2.52), we can replace u_2 in (2.54) by ω and treat r_1 as a function of n, u and u_1. Then

$$\begin{aligned}
\frac{\partial r_1}{\partial u_1} &= \frac{\omega_{,2}}{Q(n + 2, \omega)} - \frac{1}{Q(n + 1, u_1)} = -\frac{\omega_{,1} Q(n, u)}{Q(n + 1, u_1) Q(n + 2, \omega)} \\
&= -\frac{Q(n, u)}{Q(n + 1, u_1)} \frac{\partial r_1}{\partial u}.
\end{aligned} \tag{2.55}$$

(The second of these equalities is a consequence of the LSC.) If we now regard r_1 as a function of n, s and s_1, (2.55) amounts to

$$\frac{\partial r_1}{\partial s_1} + \frac{\partial r_1}{\partial s} = 0,$$

and so r_1 depends on n and $s_1 - s$ only. We have therefore reduced the OΔE (2.52) to a first-order OΔE, which is necessarily in forward form:

$$r_1 = F(n, r). \tag{2.56}$$

If (2.56) can be solved, we can go further; if the general solution is

$$r = f(n; c_1),$$

then the general solution of the OΔE (2.52) is

$$\int \frac{du}{Q(n, u)} = s = \sigma_k\{f(k; c_1); n_0, n\} + c_2. \tag{2.57}$$

As s is compatible, this solution can be inverted (in principle) to yield $u(n)$.

Example 2.19 The Lie point symmetries of

$$u_2 = \frac{2uu_1}{u_1 + u} \tag{2.58}$$

include symmetries whose characteristic is $Q(n, u) = u^2$. Use this result to reduce (2.58) to a first-order OΔE, and hence find the general solution of (2.58).

Solution: From (2.53), let

$$r = \int \frac{du_1}{(u_1)^2} - \int \frac{du}{u^2} = \frac{1}{u} - \frac{1}{u_1}.$$

(It is obvious that $s = -1/u$ is compatible.) Then, on solutions of (2.58),

$$r_1 = \frac{1}{u_1} - \frac{u_1 + u}{2uu_1} = \frac{1}{2u_1} - \frac{1}{2u} = -\frac{r}{2}.$$

The general solution of this OΔE is $r = c_1(-2)^{-n}$, so

$$-\frac{1}{u} = c_1\sigma_k\{(-2)^{-k}; 0, n\} + c_2 = \frac{c_1}{3}\left(1 - (-2)^{-n}\right) + c_2.$$

Rearranging this, we obtain the general solution of (2.58):

$$u = \frac{1}{\tilde{c}_1(-2)^{-n} + \tilde{c}_2}. \qquad \blacktriangle$$

Example 2.20 The above reduction process is a generalization of the method of reduction of order for linear OΔEs. To see this, consider the linear OΔE

$$u_2 + a_1(n)u_1 + a_0(n)u = b(n). \tag{2.59}$$

If $u = f(n)$ is any nonzero solution of the associated homogeneous equation,

$$u_2 + a_1(n)u_1 + a_0(n)u = 0,$$

then $Q(n, u) = f(n)$ is a characteristic of Lie point symmetries of (2.59), a result that also applies to higher-order linear OΔEs (see Exercise 2.6). Let

$$r = \frac{u_1}{f(n + 1)} - \frac{u}{f(n)},$$

in accordance with (2.53). Then (2.59) reduces to

$$r_1 = \frac{b(n)}{f(n+2)} + \frac{a_0(n)f(n)}{f(n+2)} r,$$

which is equivalent to (1.42), with r replacing w.   ▲

Although the above method can always be used for reduction of order, the calculation may be simpler if we replace r by a function of r. Let

$$\tilde{r} = g(r),$$

where g is a bijection[10]. Then the reduced OΔE (2.56) is equivalent to

$$\tilde{r}_1 = g\big(F(n, g^{-1}(\tilde{r}))\big).$$

Example 2.21   The characteristics of Lie point symmetries of

$$u_2 = \frac{2u_1 - u + u(u_1)^2}{1 - (u_1)^2 + 2uu_1} \tag{2.60}$$

include $Q = u^2 + 1$. Use reduction of order to solve (2.60).

Solution: We could use (2.53) to write down a first-order OΔE for

$$r = \tan^{-1}(u_1) - \tan^{-1}(u), \tag{2.61}$$

where $\tan^{-1}$ gives values in the range $(-\pi/2, \pi/2)$. However, the OΔE (2.60) is a rational polynomial, so it is convenient instead to find an OΔE for

$$\tilde{r} = \tan(r) = \frac{u_1 - u}{1 + uu_1}.$$

For this change of variables to be a bijection, we shall assume for now that $r \in (-\pi/2, \pi/2)$. A short calculation shows that

$$\tilde{r}_1 = \tilde{r}, \tag{2.62}$$

and so $\tilde{r}$ is constant; for convenience, let $\tilde{r} = \tan c_1$, where $c_1 \in (-\pi/2, \pi/2)$. Then, by assumption, $r = c_1$ and so the solution of (2.61) is

$$\tan^{-1} u = c_1 n + c_2. \tag{2.63}$$

Now let us drop our assumption on r; by definition, $r \in (-\pi, \pi)$. The reduced equation (2.62) still applies, but now r may differ from c_1 by $\pm\pi$. However, the

---

[10] It may be helpful to restrict the domain of g temporarily, checking later whether the resulting solution of the OΔE (2.52) holds for all r (see Example 2.21). More generally, one can allow $\tilde{r}$ to be any function of r and n that can be inverted to obtain r.

only effect of this is to add an integer multiple of π to the right-hand side of (2.63), which does not change u. Thus, the general solution of (2.60) is

$$u = \tan(c_1 n + c_2).$$ ▲

Example 2.22 Every linear homogeneous second-order OΔE,

$$u_2 + a_1(n)\, u_1 + a_0(n)\, u = 0, \qquad\qquad (2.64)$$

has a characteristic $Q(n, u) = u$. Use this to reduce (2.64) to a Riccati equation.

Solution: With $s = \mathrm{Log}(u)$, we obtain

$$r = \mathrm{Log}(u_1) - \mathrm{Log}(u).$$

As (2.64) is polynomial, it is convenient to choose

$$\tilde{r} = \exp(r) = u_1/u.$$

Then the reduced OΔE is

$$\tilde{r}_1 = -a_1(n) - a_0(n)/\tilde{r},$$

which is equivalent to the Riccati equation (1.85). ▲

Examples 2.20 and 2.22 demonstrate that if a particular characteristic is common to a class of OΔEs, every equation in the class can be reduced by the same r (subject to the canonical coordinate s being compatible with the OΔE). This is also true for ODEs; most methods for solving a particular class of ODEs exploit a Lie symmetry group that is shared by all ODEs in that class.

Reduction of order is not restricted to second-order OΔEs. The same process reduces a pth-order OΔE for u with a known nonzero characteristic $Q(n, u)$ to an OΔE of order $p - 1$ for $r = s_1 - s$, where s is any compatible canonical coordinate. It is easy to check that the LSC yields

$$\frac{\partial r_{p-1}}{\partial s_{p-1}} + \cdots + \frac{\partial r_{p-1}}{\partial s_1} + \frac{\partial r_{p-1}}{\partial s} = 0,$$

and consequently there exists a function F such that

$$r_{p-1} = F(n, s_1 - s, \ldots, s_{p-1} - s_{p-2}) = F(n, r, \ldots, r_{p-2}). \qquad (2.65)$$

If the general solution of (2.65) is

$$r = f(n; c_1, \ldots, c_{p-1}),$$

the general solution of the original p^{th}-order OΔE is

$$\int \frac{du}{Q(n, u)} = s = \sigma_k\{f(k; c_1, \ldots, c_{p-1}); n_0, n\} + c_p.$$

As s is compatible, this solution defines $u(n)$ uniquely (for given c_i).

2.6 First integrals: the basics

For ODEs, there is another approach to reduction of order, which is to look for first integrals[11]. This approach is equally applicable to OΔEs.

Roughly speaking, a *first integral* of an OΔE is a non-constant function that is constant on each solution of the OΔE. More precisely, suppose that

$$u_p = \omega(n, u, \ldots, u_{p-1}) \tag{2.66}$$

is defined on a regular domain D. A first integral of the OΔE (2.66) is a function $\phi(n, u, \ldots, u_{p-1})$ that is not constant but satisfies, for each $n \in D$,

$$S_n \phi = \phi \qquad \text{when (2.66) holds.} \tag{2.67}$$

If $n + 1 \in D$, we may replace n by $n + 1$ in (2.67) to obtain

$$S_n^2 \phi = S_n \phi = \phi \qquad \text{when (2.66) holds.}$$

Similarly,

$$S_n^k \phi = \phi \qquad \text{when (2.66) holds,}$$

provided that $n + k \in D$. Consequently, on solutions of (2.66), every possible shift of n leaves the value of ϕ unchanged.

The *determining equation* for first integrals is, from (2.67),

$$\phi(n + 1, u_1, \ldots, u_{p-1}, \omega(n, u, \ldots, u_{p-1})) = \phi(n, u, \ldots, u_{p-2}, u_{p-1}); \tag{2.68}$$

this must hold identically for each possible choice of $(n, u, \ldots, u_{p-1})$. For brevity, we define the *restricted shift operator*, $\overline{S}_n$, to be S_n followed by the replacement of u_p by $\omega(n, u, \ldots, u_{p-1})$. Then the determining equation (2.68) amounts to

$$(\overline{S}_n - I)\, \phi(n, u, \ldots, u_{p-2}, u_{p-1}) = 0. \tag{2.69}$$

Thus, a first integral is a non-constant element of the kernel of $\overline{S}_n - I$.

Suppose that the general solution of a given OΔE (2.66) is

$$u = f(n; c_1, \ldots, c_p).$$

The $(p-1)^{\text{th}}$ prolongation of this solution is

$$u_k = f(n + k; c_1, \ldots, c_p), \qquad k = 0, \ldots, p - 1, \tag{2.70}$$

using our convention that $u_0 = u$. We can obtain first integrals by solving the set of simultaneous equations (2.70) for each c_i, to get

$$\phi^i(n, u, \ldots, u_{p-1}) = c_i, \qquad i = 1, \ldots, p. \tag{2.71}$$

---

[11] In essence, this is dual to Lie's method – see Bluman and Anco (2002).

This associates the pth-order scalar OΔE (2.66) with a particularly simple system of first-order OΔEs:

$$(S_n - I) \phi^i = 0, \qquad i = 1, \dots, p. \qquad (2.72)$$

Example 2.23 Find all first integrals of

$$u_2 - 2u_1 + u = 0.$$

Solution: The general solution of the OΔE, prolonged to first order, gives the system

$$u = c_1 n + c_2, \qquad u_1 = c_1(n + 1) + c_2.$$

Solving this pair of equations for c_1 and c_2 yields two first integrals:

$$\phi^1(n, u, u_1) = u_1 - u, \qquad \phi^2(n, u, u_1) = (n + 1)u - nu_1.$$

We could have written the general solution in a different form, such as

$$u = \tilde{c}_1(1 + n) + \tilde{c}_2(1 - n),$$

which would have led us to different first integrals. However, whichever arbitrary constants $\tilde{c}_1, \tilde{c}_2$ are used will be (independent) functions of c_1 and c_2:

$$\tilde{c}_1 = f(c_1, c_2), \qquad \tilde{c}_2 = g(c_1, c_2).$$

On solutions of the OΔE, we can replace each c_i by ϕ^i, so the first integrals corresponding to $\tilde{c}_1$ and $\tilde{c}_2$ are

$$\tilde{\phi}^1 = f(\phi^1, \phi^2), \qquad \tilde{\phi}^2 = g(\phi^1, \phi^2).$$

Consequently, every first integral can be expressed in terms of ϕ^1 and ϕ^2. ▲

Note: It is important to recognize that, in general, (2.66) and (2.72) are not equivalent. The general solution of (2.72) is (2.71), where each c_i is entirely arbitrary, but the corresponding constants in the general solution of (2.66) may be constrained (to avoid singularities, for instance). Furthermore, (2.66) might have solutions that are not part of the general solution and do not satisfy (2.72). Example 2.8 illustrates both of these features; the OΔE has a solution, $u = 0$, that is not part of the general solution, $u = 2/(n(n - 1) - 2c_1)$. To avoid singularities in the general solution, c_1 cannot be $n_1(n_1 - 1)/2$, for any $n_1 \in \mathbb{N}$.

The first integrals $\phi^1, \dots, \phi^p$ in (2.71) constitute a *basis* for the set of first integrals of (2.66); every first integral $\tilde{\phi}$ can be expressed in terms of the basis. Finding a basis for the set of first integrals enables us to determine the general solution of the OΔE, so we need to be able to identify when p given first integrals, $\tilde{\phi}^i(n, u, \dots, u_{p-1})$, $i = 1, \dots, p$, constitute a basis. The key to doing this is temporarily to regard each $\tilde{\phi}^i$ as a function of the (unknown) basis $\phi^1, \ldots, \phi^p$. Suppose that all of these functions are differentiable (at least, locally). Then we can form two Jacobian determinants, one with respect to the variables u^i,

$$J(\tilde{\phi}^1, \tilde{\phi}^2, \ldots, \tilde{\phi}^p) = \det \begin{bmatrix} \tilde{\phi}^1_{,1} & \tilde{\phi}^1_{,2} & \cdots & \tilde{\phi}^1_{,p} \\ \vdots & \vdots & & \vdots \\ \tilde{\phi}^p_{,1} & \tilde{\phi}^p_{,2} & \cdots & \tilde{\phi}^p_{,p} \end{bmatrix}, \tag{2.73}$$

and the other with respect to the unknown basis ϕ^j,

$$K = \det \begin{bmatrix} \partial\tilde{\phi}^1/\partial\phi^1 & \partial\tilde{\phi}^1/\partial\phi^2 & \cdots & \partial\tilde{\phi}^1/\partial\phi^p \\ \vdots & \vdots & & \vdots \\ \partial\tilde{\phi}^p/\partial\phi^1 & \partial\tilde{\phi}^p/\partial\phi^2 & \cdots & \partial\tilde{\phi}^p/\partial\phi^p \end{bmatrix}.$$

By the chain rule,

$$J(\tilde{\phi}^1, \tilde{\phi}^2, \ldots, \tilde{\phi}^p) = K J(\phi^1, \phi^2, \ldots, \phi^p);$$

thus, if (2.73) is nonzero, so is K.

Any p continuous functions, $f^1, \ldots, f^p$, of a given set of p continuous variables, $\mathbf{z} = (z^1, \ldots, z^p)$, are said to be *functionally independent* if[12]

$$F(f^1(\mathbf{z}), \ldots, f^p(\mathbf{z})) = 0 \qquad \text{implies} \qquad F \equiv 0. \tag{2.74}$$

(Here F is assumed to be continuous.) The first integrals $\tilde{\phi}^i$ form a basis if and only if they are functionally independent, so that

$$F(\tilde{\phi}^1(\phi^1, \ldots, \phi^p), \ldots, \tilde{\phi}^p(\phi^1, \ldots, \phi^p)) = 0 \quad \text{implies} \quad F \equiv 0. \tag{2.75}$$

Provided that F is (locally) differentiable with respect to each of its arguments, the condition (2.75) is satisfied whenever K is nonzero. As we are avoiding any pathological cases, we have a simple test: if the Jacobian determinant (2.73) is nonzero, the set $\tilde{\phi}^1, \ldots, \tilde{\phi}^p$ is a basis.

It is rarely obvious how to obtain p independent first integrals of a given OΔE. Nevertheless, it is possible to make progress incrementally. If we can find even one first integral, ϕ^1, all that remains is to solve the reduced OΔE,

$$\phi^1(n, u, \ldots, u_{p-1}) = c_1. \tag{2.76}$$

It is easy to check that ϕ^1 necessarily depends on both u and u_{p-1}, so the reduced OΔE (2.76) is genuinely of order $p-1$. If we can find two first integrals

---

[12] This definition is inadequate if $\mathbf{z}$ includes discrete variables. For instance, suppose that $\mathbf{z} = (m, n, z) \in \mathbb{Z}^2 \times \mathbb{R}$. Then $f^1(\mathbf{z}) = (-1)^m z$ and $f^2(\mathbf{z}) = (-1)^n z$ are independent (as f^1 is not a function of f^2 and vice versa), but $(f^1)^2 - (f^2)^2 = 0$. Such difficulties do not affect first integrals, for they are continuous functions of any basis, $\phi^1, \ldots, \phi^p$.

and use $\phi^1 = c_1$ and $\phi^2 = c_2$ to eliminate u_{p-1} (or u), we will obtain an OΔE of order $p - 2$, and so on. There is no guarantee that a reduced OΔE will be in forward (or backward) form. However, the solution of the original OΔE is unique for a given set of initial conditions; this enables us to eliminate any spurious solutions that occur in the general solution of the reduced OΔE (see Exercise 2.11).

Even if the reduced OΔE can only be solved for certain particular values of the constants c_i, this will still yield some families of solutions of (2.66). So it is a good idea to try to find as many functionally independent first integrals as possible.

2.7 Direct methods for finding first integrals

The simplest way to find first integrals of a given OΔE is by inspection. We need to find a function $\phi(n, u, \ldots, u_{p-1})$ that depends on u and u_{p-1} and satisfies the condition (2.67). It may be fairly obvious how to rearrange the OΔE to achieve this.

Example 2.24 Find a first integral of

$$u_2 = (u_1)^2/u.$$

Solution: When we rewrite the OΔE as

$$u_2/u_1 = u_1/u,$$

it becomes obvious that $\phi^1(n, u, u_1) = u_1/u$ is a first integral. In this very simple example, we can go on to solve the reduced OΔE, which amounts to $u_1 = c_1 u$. The general solution, prolonged to first order, is

$$u = c_2(c_1)^n, \qquad u_1 = c_2(c_1)^{n+1}.$$

Therefore a second independent first integral is

$$\phi^2 = u^{n+1}/(u_1)^n,$$

which is not particularly obvious. ▲

Example 2.25 Consider the OΔE

$$u_3 = \frac{u_1(u_2 + 1)}{u + 1}.$$

This can be rearranged as

$$\left(S_n^2 - 1\right) \frac{u_1}{u + 1} = 0, \tag{2.77}$$

so the OΔE has the first integral

$$\phi^1 = (S_n + I)\frac{u_1}{u+1} = \frac{u_2}{u_1 + 1} + \frac{u_1}{u+1}.$$

One can go further, by solving (2.77) to obtain the first-order linear OΔE

$$u_1 = (c_1 + c_2(-1)^n)(u + 1),$$

which can be solved by the methods of Chapter 1. ▲

If inspection fails, one can adopt the same strategy that is used to obtain symmetries: select an ansatz for ϕ, then use differential elimination to reduce the determining equation (2.68) to a differential equation.

For second-order OΔEs, for instance, ϕ must depend upon both u and u_1. The simplest ansatz is to assume that ϕ is linear in these variables, but this works only when the OΔE is linear. For nonlinear OΔEs, it may be useful to assume that ϕ is linear in one of the continuous variables; the dependence on the other one can be found from the determining equation (by differential elimination).

Example 2.26 Find a first integral of the OΔE

$$u_2 = u - \frac{2u_1}{u_1 + n}.$$

Solution: Let us try an ansatz that is linear in u_1:

$$\phi(n, u, u_1) = f(n, u) u_1 + g(n, u). \tag{2.78}$$

The determining equation amounts to

$$f(n+1, u_1)\left(u - \frac{2u_1}{u_1 + n}\right) + g(n+1, u_1) = f(n, u) u_1 + g(n, u). \tag{2.79}$$

Differentiating (2.79) with respect to u gives

$$f(n+1, u_1) = f'(n, u) u_1 + g'(n, u). \tag{2.80}$$

Now differentiate (2.80) with respect to u_1, to obtain

$$f'(n+1, u_1) = f'(n, u),$$

the general solution of which is

$$f(n, u) = cu + \beta(n),$$

where c is an arbitrary constant. Thus, the general solution of (2.80) is

$$g(n, u) = \beta(n+1) u + \gamma(n).$$

Finally, substitute f and g into the determining equation (2.79) and compare powers of u_1. This establishes that the only solutions (2.78) are multiples of

$$\phi(n, u, u_1) = (u + n - 1)u_1 + nu.$$

The reduced OΔE, $\phi = c_1$, is a Riccati equation. The solution is particularly simple when $c_1 = 0$:

$$u = (n - 1)\left(c_2^{(-1)^n} - 1\right), \qquad c_2 \neq 0.$$

(The case $c_1 \neq 0$ is left to the reader.)   ▲

For OΔEs of order p, any ansatz for ϕ must depend on both u and u_{p-1}. Choosing a good ansatz is something of an art. A very restrictive ansatz is easy to check, but is unlikely to yield solutions of (2.68). Trying a very general ansatz spreads the search much wider, but the computations will be lengthy and may even become intractable.

2.8 Indirect methods for constructing first integrals

So far, we have attempted to find first integrals directly from the OΔE or the determining equation. However, it is often more productive to take an indirect approach, as follows.

Suppose that ϕ is a first integral of the OΔE (2.66) and that Q is a characteristic of a one-parameter Lie group of symmetries. Let

$$\tilde{\phi} = \overline{X}\phi,$$

where

$$\overline{X} = Q\frac{\partial}{\partial u} + \overline{S}_n Q\frac{\partial}{\partial u_1} + \cdots + \overline{S}_n^{p-1} Q\frac{\partial}{\partial u_{p-1}}. \tag{2.81}$$

It is helpful to introduce the functions

$$P_i(n, u, \ldots, u_{p-1}) = \phi_{,i}(n, u, \ldots, u_{p-1}), \tag{2.82}$$

so that

$$\tilde{\phi} = QP_1 + (\overline{S}_n Q)P_2 + \cdots + (\overline{S}_n^{p-1} Q)P_p. \tag{2.83}$$

The functions P_i satisfy a system that is obtained by differentiating the determining equation (2.68) with respect to each u_i in turn:

$$\omega_{,1}\,\overline{S}_n P_p = P_1,$$
$$\overline{S}_n P_i + \omega_{,i+1}\,\overline{S}_n P_p = P_{i+1}, \qquad i = 1, \ldots, p - 1. \tag{2.84}$$

These identities, together with the linearized symmetry condition,

$$\overline{S}_n^p Q = \omega_{,p} \overline{S}_n^{p-1} Q + \cdots + \omega_{,1} Q, \tag{2.85}$$

establish that $\tilde{\phi}$ is either a first integral or a constant, because

$$\overline{S}_n \tilde{\phi} = (\overline{S}_n Q)(\overline{S}_n P_1) + (\overline{S}_n^2 Q)(\overline{S}_n P_2) + \cdots + (\omega_{,p} \overline{S}_n^{p-1} Q + \cdots + \omega_{,1} Q)(\overline{S}_n P_p)$$
$$= \tilde{\phi}.$$

This observation raises the possibility of using symmetries to create new first integrals from known ones. Of course there is no guarantee that $\tilde{\phi}$ and ϕ will be functionally independent. Nevertheless, it is very easy to calculate $\tilde{\phi}$ and find out whether it is new.

Example 2.27   The OΔE

$$u_3 = u_2 u_1 / u$$

can be written as

$$(S_n^2 - I)(u_1/u) = (S_n - I)(S_n + I)(u_1/u) = 0.$$

Thus $\phi^1 = u_2/u_1 + u_1/u$ is a first integral. The corresponding P_i are

$$P_1 = -u_1/u^2, \qquad P_2 = 1/u - u_2/(u_1)^2, \qquad P_3 = 1/u_1.$$

All characteristics of Lie point symmetries of the OΔE are linear combinations of

$$Q_1 = u, \qquad Q_2 = (-1)^n u, \qquad Q_3 = nu, \qquad Q_4 = u \ln|u|.$$

By substituting each Q_i in turn into (2.83), we obtain the corresponding $\tilde{\phi}^i$:

$$\tilde{\phi}^1 = 0;$$
$$\tilde{\phi}^2 = 2(-1)^n (u_2/u_1 - u_1/u);$$
$$\tilde{\phi}^3 = u_2/u_1 + u_1/u;$$
$$\tilde{\phi}^4 = (u_1/u) \ln|u_1/u| + (u_2/u_1) \ln|u_2/u_1|.$$

Clearly, $\tilde{\phi}^2$ is independent of ϕ^1. The remaining first integrals give nothing new. (It is easy to check that the Jacobian determinant $J(\phi^1, \tilde{\phi}^2, \tilde{\phi}^4)$ is zero, though some work is needed to establish how $\tilde{\phi}^4$ depends on ϕ^1 and $\tilde{\phi}^2$.) However, ϕ^1 and $\tilde{\phi}^2$ can be used to write u_1/u as a function of n and two arbitrary constants; this amounts to a first-order linear OΔE that is easy to solve.   ▲

We obtained $\overline{S}_n \tilde{\phi} = \tilde{\phi}$ from the system (2.84), which is a necessary but not sufficient set of conditions for the determining equation to be satisfied. Any solution of this system, $(P_1, \ldots, P_p)$, can be combined with symmetries to create first integrals (or constants). Generally speaking, not every solution satisfies the integrability conditions

$$P_{i,j} = P_{j,i}, \qquad i > j, \tag{2.86}$$

that are necessary for the existence of a function ϕ that satisfies (2.82). However, any solution that does satisfy the integrability conditions (2.86) can be used independently of the symmetries to construct the first integral

$$\phi = \int P_1 \, du + P_2 \, du_1 + \cdots + P_p \, du_{p-1} + \alpha(n). \tag{2.87}$$

Here $\alpha(n)$ is found from the determining equation $\overline{S}_n \phi = \phi$.

In summary, given a solution of (2.84), one can always combine it with symmetries to create first integrals. If the solution satisfies the integrability conditions (2.86), there is an extra first integral, (2.87), which might be independent of any that were found using symmetries. Consequently, this indirect method can be very fruitful.

To solve the system (2.84) by differential elimination, one needs an ansatz for P_p. It is helpful to eliminate the remaining functions P_i; this produces the following scalar determining equation for P_p:

$$\overline{S}_n^{p-1}(\omega_{,1} \overline{S}_n P_p) + \overline{S}_n^{p-2}(\omega_{,2} \overline{S}_n P_p) + \cdots + \omega_{,p} \overline{S}_n P_p - P_p = 0. \tag{2.88}$$

As always, the choice of ansatz is something of an art. For instance, to find all solutions for which ϕ is linear in u_{p-1}, look for solutions P_p that are independent of u_{p-1}.

If the OΔE is of the form

$$u_p = \prod_{i=0}^{p-1} a_i(n, u_i), \tag{2.89}$$

for some functions a_i, the ansatz

$$P_p = v(n)/u_{p-1} \tag{2.90}$$

reduces the determining equation for P_p to

$$\frac{u_{p-1} \, a_0'(n+p-1, u_{p-1})}{a_0(n+p-1, u_{p-1})} v_p + \cdots + \frac{u_{p-1} \, a_{p-1}'(n, u_{p-1})}{a_{p-1}(n, u_{p-1})} v_1 - v_0 = 0. \tag{2.91}$$

(The ansatz (2.90) has a nice property: the functions P_i, which one reconstructs from (2.84), automatically satisfy the integrability conditions (2.86).) If $a_i(n, u_i) = u_i^{\alpha_i(n)}$ for each i, where $\alpha_i(n)$ is given, the reduced equation (2.91) is an OΔE. Otherwise, it yields an overdetermined system of OΔEs, which is obtained by gathering terms according to their dependence on u_{p-1}.

Example 2.28   Consider the third-order OΔE

$$u_3 = \frac{1}{u}(u_2/u_1)^{(-1)^n}, \qquad n \geq 0, \tag{2.92}$$

which is of the form (2.89). For simplicity, suppose that $u(0)$, $u(1)$ and $u(2)$ are all positive, so that $u > 0$ for every $n \geq 0$.

With the ansatz $P_3 = v(n)/u_2$, the reduced equation (2.91) is the OΔE

$$-v_3 + (-1)^n v_2 + (-1)^n v_1 - v = 0,$$

which has an obvious nonzero solution, $v = (-1)^n$. By the method of reduction of order, we obtain the following OΔE for $w = (S_n - I)((-1)^n v)$:

$$w_2 + \{1 + (-1)^n\}w_1 + w = 0.$$

The general solution of this OΔE is

$$w = 2c_1 \cos\left(\frac{n\pi}{2}\right) + 2c_2\left(\sin\left(\frac{n\pi}{2}\right) + n\cos\left(\frac{n\pi}{2}\right)\right),$$

so

$$v = (-1)^n\left\{c_1\left(\sin\left(\frac{n\pi}{2}\right) - \cos\left(\frac{n\pi}{2}\right)\right) + c_2\left((n-1)\sin\left(\frac{n\pi}{2}\right)\right.\right.$$
$$\left.\left. - n\cos\left(\frac{n\pi}{2}\right)\right) + c_3\right\}.$$

From here, it is straightforward to reconstruct each P_i and thus to derive three functionally independent first integrals:

$$\phi^1 = (-1)^n\left(\cos\left(\frac{n\pi}{2}\right)\ln(u/u_2) + \sin\left(\frac{n\pi}{2}\right)\ln(uu_2)\right);$$

$$\phi^2 = (-1)^n\left\{\cos\left(\frac{n\pi}{2}\right)(n\ln(u/u_2) - 2\ln(u_1)) + \sin\left(\frac{n\pi}{2}\right)(n\ln(uu_2) + \ln(u/u_2))\right\};$$

$$\phi^3 = (1 + (-1)^{n+1})\ln(u_1) + (-1)^n\ln(uu_2).$$

Setting each of these to be an arbitrary constant yields the general solution of the OΔE (2.92), which (after simplification) is

$$u = \exp\left\{\tilde{c}_1\left(\cos\left(\frac{n\pi}{2}\right) + (n-1)\sin\left(\frac{n\pi}{2}\right)\right) + \tilde{c}_2\sin\left(\frac{n\pi}{2}\right) + \tilde{c}_3\cos^2\left(\frac{n\pi}{2}\right)\right\}. \quad \blacktriangle$$

Notes and further reading

Sophus Lie established the foundations of symmetry analysis towards the end of the 19th century. One of his aims was to unify existing solution methods for differential equations. Lie groups and Lie algebras remained popular research areas throughout the first half of the 20th century, but symmetry methods were largely forgotten until the resurgence of interest in symmetry analysis in recent decades. The wonderfully comprehensive text by Olver (1993) discusses the reasons for this. For a simple introduction to symmetry methods, see Hydon (2000b) or Stephani (1989). The text by Bluman and Anco (2002) is more detailed, with many worked examples and a particular emphasis on first integrals of ODEs.

Surprisingly, symmetry analysis of OΔEs was not introduced until recently. Maeda (1987) first stated the linearized symmetry condition (in a restricted form) and used ansätze to obtain some simple solutions. Later, series-based solution methods were developed by Quispel and Sahadevan (1993), Gaeta (1993) and Levi et al. (1997). Differential elimination (also called 'invariant differentiation' – see Hydon, 2000a) has become the standard method for obtaining closed-form solutions of the linearized symmetry condition. This approach has been given a rigorous foundation, using differential algebra, by Mansfield and Szanto (2003).

A natural question arises: can one construct finite difference approximations that preserve the symmetries of a given ODE? This is one theme within geometric integration, which aims to create numerical methods that preserve geometrical and topological features of a given system. The review by Budd and Piggott (2003) is a very accessible introduction to some key concepts in geometric integration; for a more detailed survey, see Hairer et al. (2006). There are various approaches to symmetry preservation; some are general (see Dorodnitsyn, 2001, 2011; Kim and Olver, 2004) while others deal with special types of ODEs (see McLachlan and Quispel, 2001; Iserles et al., 2000; Leimkuhler and Reich, 2004).

First integrals have a key role in the theory of discrete integrable systems, and particularly in the study of integrable maps. The most widely used class of such maps (QRT) was introduced in Quispel et al. (1988); it depends on 18 parameters. Each QRT map amounts to a second-order OΔE, with a first integral that is a ratio of biquadratic polynomials in u and u_1. The properties of QRT maps are described in detail in Duistermaat (2010). Some simple types of QRT map (the Lyness equation and the McMillan map) are examined in Exercises 2.11 and 2.16 below.

Exercises

2.1   Explain why the following initial-value problems cannot be solved:

(a) $u_1 = 1/(u - 1)$, $n \geq 0$, $u(0) = 8/5$;

(b) $u_1 = 2/(2 - u - |u|)$, $n \geq 0$, $u(0) = a \in \mathbb{R}$.

2.2   Find a characteristic $Q(n, u)$ for

$$u_1 = (nu - 1)/(u + n),$$

and hence solve this OΔE.

2.3   Show that every scalar Riccati OΔE ,

$$u_1 = \frac{a_2(n)\,u + a_3(n)}{u + a_1(n)}, \qquad a_3(n) \neq a_1(n)\,a_2(n),$$

has three linearly independent characteristics that are quadratic in u. Find the characteristics when each $a_i(n)$ is constant.

2.4   Use a suitable ansatz to find a characteristic for

$$u_1 = \sqrt{u^2 + n^2}, \qquad n \geq 0.$$

Hence calculate the general solution of this OΔE.

2.5   Solve each of the following OΔEs by finding all characteristics of Lie point symmetries and using them for reduction of order:

(a) $u_2 = e^{-n} u^2 u_1$;
(b) $u_2 = (u_1/u)^{n+2}, \quad n \geq 1$;
(c) $u_2 = \{u_1(u_1 - u) + u\}/\{u_1 - u + 1\}$;
(d) $u_2 = \{2u(u_1)^2 + u_1 + u\}/\{(u_1)^2 + uu_1 + 2\}$;
(e) $u_2 = a(n)(u_1)^2/u + b(n)\,u_1, \qquad a(n) \neq 0, \quad b(n) \neq 0.$

[Write a short computer algebra program to do the calculations.]

2.6   Show that every linear homogeneous OΔE,

$$u_p + a_{p-1}(n)\,u_{p-1} + \cdots + a_0(n)\,u = 0, \qquad n \in D,$$

has characteristics of the form

$$Q(n, u) = c_1 u + \beta(n),$$

where $u = \beta(n)$ is any solution of the OΔE. Now prove that if $p \geq 2$ and $a_{p-1}(n)$ is nonzero for each $n \in D$, *every* characteristic of Lie point symmetries is of the above form.

2.7   This question explores the linearization of a second-order OΔE,

$$u_2 = \omega(n, u, u_1), \tag{2.93}$$

by a transformation of the form

$$\Psi : \quad (n, u) \mapsto (n, \hat{u}) = (n, \psi_n(u)), \qquad \psi_n'(u) \neq 0. \tag{2.94}$$

(a) Solve $\eta_{,12} = 0$, to find all functions ω for which (2.43) is an ODE. Use this result to conclude that if the OΔE (2.93) is linearizable then $\eta_{,12} = 0$.

(b) Show that if (2.93) can be linearized to an inhomogeneous OΔE by a transformation (2.94), it can be linearized to a homogeneous OΔE:

$$\hat{u}_2 + a_1(n)\hat{u}_1 + a_0(n)\hat{u} = 0.$$

(c) Show how the characteristics of (2.93) can be used to construct a linearizing transformation.

2.8 (a) Show that if $Q(n, u)$ is a characteristic for

$$u_4 = \omega(u, u_2), \tag{2.95}$$

$(-1)^n Q(n, u)$ is also a characteristic for this OΔE.

(b) How are the characteristics of Lie point symmetries of (2.95) related to those of $u_2 = \omega(u, u_1)$?

(c) How do the answers above change if the right-hand side of (2.95) is replaced by $\omega(n, u, u_2)$?

2.9 (a) Determine the general solution of the OΔE (2.48).

(b) Find another one-parameter family of solutions of (2.48) that is not included in the general solution.

(c) Determine the solution of (2.48) that satisfies the initial conditions $u(0) = 1, u(1) = -1$.

2.10 Let $\phi(n, u)$ be a first integral of $u_1 = \omega(n, u)$. Show that the LSC amounts to

$$\phi'(n + 1, \omega(n, u))Q(n + 1, \omega(n, u)) = \phi'(n, u)Q(n, u).$$

Use this result to show that the general solution of the LSC is

$$Q(n, u) = F(\phi(n, u))/\phi'(n, u),$$

where F is an arbitrary function (see Example 2.12).

2.11 (a) Show that the *Lyness equation*,

$$u_2 = (c + u_1)/u, \qquad c > 0, \tag{2.96}$$

has the first integral

$$\phi = (c + u_1 + u)(1 + 1/u_1)(1 + 1/u).$$

Use this result to find a first integral[13] of *Todd's equation*,

$$u_3 = (c + u_2 + u_1)/u, \qquad c > 0.$$

(b) Let $c = 1$. Find every solution $u(n), n \geq 0$, of the Lyness equation
(2.96) that satisfies the pair of initial conditions $u(0) = u(1) = 1$.
Do the same for the reduced OΔE, $\phi = 12$. [Hint: calculate $(S_n - I)\phi$
explicitly.] What do you observe?

2.12 Find a first integral of the OΔE

$$u_2 = -(1 + u)(1 + 1/u_1).$$

[Try inspection first; if this fails, use an ansatz for ϕ or P_2.] Hence determine the general solution of the OΔE.

2.13 Show that the OΔE

$$u_2 = \frac{1}{4}\left(u_1 + 5u + 3\left\{1 + (u_1 + u)^2\right\}^{1/2}\right)$$

has a first integral of the form $\phi = f(n, u_1 + u)$. Hence solve the OΔE.

2.14 Use the ansatz (2.90) to find first integrals of

$$u_2 = (u_1)^4/u^4,$$

and use these to write down the OΔE's general solution.

2.15 Calculate all characteristics $Q(n, u)$ for Lie point symmetries of

$$u_2 = nu/(n + 1) + 1/u_1 \qquad n \geq 1.$$

Now use an ansatz for P_2 to find as many first integrals of the OΔE as
you can. Hence find the general solution of the OΔE.

2.16 (a) Use an ansatz for P_2 to determine a first integral of

$$u_2 = -u - u_1 + 1/u_1.$$

(b) Extend your result to the *McMillan map*,

$$u_2 = -u - \frac{c_2(u_1)^2 + c_4 u_1 + c_5}{c_1(u_1)^2 + c_2 u_1 + c_3}.$$

[13] Two first integrals of Todd's equation are known – see Gao et al. (2004).

2.17 Find the conditions on the functions a, b, c and d under which

$$u_2 = d(n)\,u + \frac{a(n)\,u_1 + b(n)}{u_1 + c(n)}, \qquad a(n)c(n) - b(n) \neq 0,$$

has a first integral that is linear in u_1.

2.18 This question deals with first integrals of $u_3 = u_2 u_1 / u$. Show that neither u_1/u nor u_2/u_1 is a first integral of this OΔE, but

$$\phi^1 = f(u_1/u) + f(u_2/u_1), \qquad \phi^2 = (-1)^n \{g(u_1/u) - g(u_2/u_1)\} \quad (2.97)$$

satisfy $\overline{S}\phi^i = \phi^i$ for all functions f and g. Explain this result and show that no more than two functionally independent first integrals can be obtained from (2.97), no matter how f and g are chosen.

2.19 (a) Establish conditions under which a *nonlinear* OΔE, $u_2 = \omega(n, u, u_1)$, has a P_2 that depends on u, but not u_1.

(b) Show that if these conditions hold, one can solve

$$P_1 = \phi_{,1}(n, u, u_1), \qquad P_2 = \phi_{,2}(n, u, u_1),$$

and that the reduced OΔE, $\phi = c_1$, is a Riccati equation.

(c) Show that every scalar Riccati OΔE can be obtained in this way, by calculating the function ω that corresponds to the general Riccati equation,

$$u_1 = \frac{a(n)\,u + b(n)}{u + c(n)}, \qquad a(n)c(n) - b(n) \neq 0.$$

3

Extensions of basic symmetry methods

> We must not think of the things we could do with,
> but only of the things that we can't do without.
>
> *(Jerome K. Jerome, Three Men in a Boat)*

Chapter 2 showed that Lie point symmetries are able to simplify or solve various nonlinear scalar O∆Es. Some natural questions arise.

- Do symmetry methods also work for systems of O∆Es?
- Some ODEs have important Lie symmetries that are not point symmetries; is the same true for O∆Es?
- A nonzero characteristic reduces the order of a given O∆E by one; can R independent characteristics be used to reduce the order by R?
- Is there a connection between Lie symmetries and first integrals?

This chapter explains how to answer these questions.

3.1 Geometrical aspects of Lie point transformations

So far, we have seen various Lie point transformations that act on a single dependent variable. Everything generalizes easily to M dependent variables, $u^1, \ldots, u^M$, and it is worth taking a closer look at the geometry of such transformations. This section deals with transformations of the dependent variables only; n is suppressed (because it is fixed). A one-parameter local Lie group of point transformations acts on $\mathbb{R}^M$ by mapping each $\mathbf{u} = (u^1, \ldots, u^M)$ to a set of points,

$$\hat{\mathbf{u}}(\mathbf{u}; \varepsilon) = \mathbf{u} + \varepsilon \, \mathbf{Q}(\mathbf{u}) + O(\varepsilon^2), \tag{3.1}$$

where each Q^α is a smooth function. The M-tuple $\mathbf{Q}(\mathbf{u}) = (Q^1(\mathbf{u}), \ldots, Q^M(\mathbf{u}))$ is the *characteristic* of the transformation group.

The subset of all points $\hat{\mathbf{u}}$ that are reached from $\mathbf{u}$ by varying ε continuously from zero is called the *orbit* through $\mathbf{u}$. As ε is a single real-valued parameter,

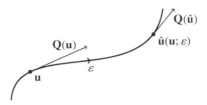

Figure 3.1 A one-dimensional orbit, showing the tangent vectors at **u** and **û**(**u**; ε).

the orbit is at most one-dimensional (see Figure 3.1). Indeed, as point transformations are smooth, the orbit is either one-dimensional or it consists only of the point **u**. The latter case occurs when **Q(u)** = **0**; then **u** is said to be an *invariant point*. Whatever the dimension of the orbit is, its tangent at **u** is

$$\frac{d\hat{\mathbf{u}}(\mathbf{u}; \varepsilon)}{d\varepsilon}\bigg|_{\varepsilon=0} = \mathbf{Q}(\mathbf{u});$$

similarly, the tangent at each **û** on the orbit is

$$\frac{d\hat{\mathbf{u}}(\mathbf{u}; \varepsilon)}{d\varepsilon} = \mathbf{Q}(\hat{\mathbf{u}}).$$

The set of tangent vectors to all orbits form a locally smooth vector field, **Q**, on $\mathbb{R}^M$. Conversely, given such a vector field, one can determine the orbits of the corresponding one-parameter local Lie group of point transformations by solving the initial-value problem

$$\frac{d\hat{\mathbf{u}}}{d\varepsilon} = \mathbf{Q}(\hat{\mathbf{u}}), \qquad \hat{\mathbf{u}}\,|_{\varepsilon=0} = \mathbf{u}. \tag{3.2}$$

Example 3.1 Find the orbits for the one-parameter local Lie group whose tangent vector field on $\mathbb{R}^2$ (with coordinates (u, v)) is

$$\mathbf{Q}(u, v) = (-v, u). \tag{3.3}$$

Solution: The initial-value problem (3.2) amounts to

$$\frac{d\hat{u}}{d\varepsilon} = -\hat{v}, \qquad \frac{d\hat{v}}{d\varepsilon} = \hat{u}, \qquad \text{subject to} \quad (\hat{u}, \hat{v})\,|_{\varepsilon=0} = (u, v). \tag{3.4}$$

The vector field is zero only at $(u, v) = (0, 0)$, which is thus an invariant point. The remaining orbits are circles centred on $(0, 0)$, as shown in Figure 3.2, because

$$\frac{d}{d\varepsilon} (\hat{u}^2 + \hat{v}^2) = 0, \qquad \text{when (3.4) holds.}$$

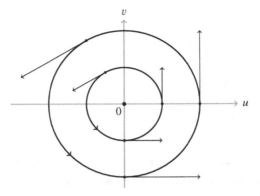

Figure 3.2 The vector field (3.3): some orbits and tangent vectors.

The solution of (3.4) away from $(0, 0)$ is easily found (locally) with the aid of polar coordinates. Substitute $\hat{u} = \hat{r}\cos\hat{\theta}$, $\hat{v} = \hat{r}\sin\hat{\theta}$ into (3.4), to obtain

$$\frac{\mathrm{d}\hat{r}}{\mathrm{d}\varepsilon} = 0, \qquad \frac{\mathrm{d}\hat{\theta}}{\mathrm{d}\varepsilon} = 1, \qquad \text{subject to} \quad (\hat{r}, \hat{\theta})|_{\varepsilon=0} = (r, \theta).$$

This leads to the following solution of (3.4):

$$(\hat{u}, \hat{v}) = (r\cos(\theta + \varepsilon), \, r\sin(\theta + \varepsilon)) = (u\cos\varepsilon - v\sin\varepsilon, \, u\sin\varepsilon + v\cos\varepsilon).$$

Although polar coordinates are not valid globally, the solution in terms of u, v and ε holds everywhere. ▲

In the above example, polar coordinates are convenient because they enable us to regard the vector field as being locally uniform, except in any neighbourhood that contains the invariant point. The same idea works for every one-parameter local Lie group. For any non-invariant point $\mathbf{u}$, there exist local canonical coordinates, $(r^1, \ldots, r^{M-1}, s)$, in terms of which the vector field is uniform in some neighbourhood of $\mathbf{u}$:

$$\hat{r}^i = r^i, \qquad \hat{s} = s + \varepsilon. \tag{3.5}$$

Here the coordinates with and without carets are evaluated at $\hat{\mathbf{u}}$ and $\mathbf{u}$, respectively. To find such coordinates, use the chain rule to obtain

$$0 = \frac{\mathrm{d}\hat{r}^i}{\mathrm{d}\varepsilon}\bigg|_{\varepsilon=0} = Q^\alpha(\mathbf{u})\frac{\partial r^i}{\partial u^\alpha}, \qquad 1 = \frac{\mathrm{d}\hat{s}}{\mathrm{d}\varepsilon}\bigg|_{\varepsilon=0} = Q^\alpha(\mathbf{u})\frac{\partial s}{\partial u^\alpha}. \tag{3.6}$$

Note: The Einstein summation convention is used from here on: if an index occurs twice in any term, once 'up' and once 'down', one sums over all possible values of that index[1]. In (3.6), each right-hand term is summed over α.

The first-order linear partial differential equations (PDEs) (3.6) can be solved for r^i and s by the method of characteristics. Any solution may be used as a local set of canonical coordinates, provided that the transformation from the coordinates $(u^1, \ldots, u^M)$ is a locally defined diffeomorphism.

The *infinitesimal generator* of the one-parameter Lie group whose characteristic (in coordinates u^α) is $\mathbf{Q}(\mathbf{u})$ is the differential operator

$$X = Q^\alpha(\mathbf{u})\,\partial_{u^\alpha}, \qquad \text{where } \partial_{u^\alpha} \equiv \frac{\partial}{\partial u^\alpha}. \tag{3.7}$$

As $Q^\alpha(\mathbf{u}) = Xu^\alpha$, the following identity holds:

$$X = (Xu^\alpha)\,\partial_{u^\alpha}. \tag{3.8}$$

In any other coordinate system, $(v^1, \ldots, v^M)$, the chain rule for derivatives gives

$$X = Q^\alpha(\mathbf{u}(\mathbf{v}))\frac{\partial v^\beta}{\partial u^\alpha}\,\partial_{v^\beta} = (Xv^\beta)\,\partial_{v^\beta}, \tag{3.9}$$

which is of the same form as (3.8). Therefore, unlike the characteristic, the infinitesimal generator is a coordinate-independent representation of the tangent vector field. In any coordinates, v^β, the set of differential operators $\partial_{v^1}, \ldots, \partial_{v^M}$ is regarded as a basis for the set of all possible tangent vector fields.

The infinitesimal generator can be used to characterize the action of the one-parameter Lie group, as follows. The PDEs (3.6) that determine canonical coordinates amount to

$$Xr^i = 0, \qquad Xs = 1, \tag{3.10}$$

and therefore $X = \partial_s$. Consider the action of the transformation group on a real analytic function, $F(\mathbf{u}) = G(\mathbf{r}(\mathbf{u}), s(\mathbf{u}))$; here $\mathbf{r} = (r^1, \ldots, r^{M-1})$. Then

$$F(\hat{\mathbf{u}}) = G(\mathbf{r}(\hat{\mathbf{u}}), s(\hat{\mathbf{u}})) = G(\mathbf{r}(\mathbf{u}), s(\mathbf{u})+\varepsilon) = \sum_{j=0}^{\infty} \frac{\varepsilon^j}{j!}\frac{\partial^j G}{\partial s^j}(\mathbf{r}(\mathbf{u}), s(\mathbf{u}))$$

$$= \sum_{j=0}^{\infty} \frac{\varepsilon^j}{j!}\,X^j G(\mathbf{r}(\mathbf{u}), s(\mathbf{u})) = \sum_{j=0}^{\infty} \frac{\varepsilon^j}{j!}\,X^j F(\mathbf{u}) = e^{\varepsilon X} F(\mathbf{u}), \tag{3.11}$$

where $e^{\varepsilon X}$ is the (formal) *exponential* of εX. In particular,

$$\hat{u}^\alpha = e^{\varepsilon X} u^\alpha, \tag{3.12}$$

---

[1] Every derivative with respect to a variable with an up (resp. down) index has that index counted as down (resp. up). For instance, $\partial s/\partial u^\alpha$ is regarded as having a single down index, α.

so the terms of all orders in ε are entirely determined by X. This proves that $\mathbf{u}$ is an invariant point if and only if $\mathbf{Q}(\mathbf{u}) = 0$. Similarly, a function $F(\mathbf{u})$ is invariant under the action of the one-parameter Lie group if and only if

$$F(\hat{\mathbf{u}}) = F(\mathbf{u})$$

for all ε sufficiently close to zero. By (3.11), $F(\mathbf{u})$ is invariant if and only if

$$XF(\mathbf{u}) = 0. \tag{3.13}$$

For instance, the canonical coordinates $r^i(\mathbf{u})$ are invariant, whereas $s(\mathbf{u})$ is not invariant.

3.2 Lie point symmetries of systems of OΔEs

We now apply the above results to a given first-order system of OΔEs for $\mathbf{u}$,

$$u_1^\alpha = \omega^\alpha(n, \mathbf{u}), \qquad \alpha = 1, \dots, M. \tag{3.14}$$

(When $M = 2$, it is convenient to use (u, v) instead of (u^1, u^2).) Any higher-order system may be reduced to a first-order system by introducing extra dependent variables, so this is the only case that we shall consider. For any transformation that fixes n, the symmetry condition is

$$\hat{u}_1^\alpha = \omega^\alpha(n, \hat{\mathbf{u}}), \qquad \alpha = 1, \dots, M, \qquad \text{when (3.14) holds.} \tag{3.15}$$

Expanding the symmetry condition to first order in ε gives the LSC for $\mathbf{Q}(n, \mathbf{u})$:

$$Q^\alpha(n + 1, \, \omega(n, \mathbf{u})) = \frac{\partial \omega^\alpha(n, \mathbf{u})}{\partial u^\beta} \, Q^\beta(n, \mathbf{u}), \qquad \alpha = 1, \dots, M. \tag{3.16}$$

Solutions of this system of functional equations are found, as in the scalar case, by choosing an ansatz for $\mathbf{Q}$ and then using differential elimination. For instance, one might assume that each Q^α is linear in $\mathbf{u}$, or that each Q^α depends on u^α and n only. Using a simple ansatz makes the LSC fairly easy to solve, but increases the risk of obtaining $\mathbf{Q}(n, \mathbf{u}) = \mathbf{0}$ as the only solution.

Having found a characteristic, one simplifies the system of OΔEs (3.14) by using compatible canonical coordinates, $(\mathbf{r}, s)$. Here, compatibility means that it is possible to describe the system of OΔEs locally in terms of the canonical coordinates, which are obtained from the original coordinates $\mathbf{u}$ by a locally defined diffeomorphism. (Just as for scalar OΔEs, one may need to restrict the domain or introduce complex-valued canonical coordinates.) Then (locally) the system (3.14) amounts to

$$r_1^i = f^i(n, \mathbf{r}), \qquad s_1 - s = g(n, \mathbf{r}). \tag{3.17}$$

The functions f^i and g are independent of s, for otherwise (3.17) would not admit the Lie symmetries generated by $X = \partial_s$.

Example 3.2 Find all Lie point symmetry generators that are of the form $X = Q^1(n, u) \partial_u + Q^2(n, v) \partial_v$ for the system of OΔEs

$$u_1 = \frac{u}{uv + 2}, \qquad v_1 = -\frac{1}{u}, \tag{3.18}$$

and hence obtain the general solution of this system.

Solution: The LSC is

$$Q^1\left(n + 1, \frac{u}{uv + 2}\right) = \frac{2}{(uv + 2)^2} Q^1(n, u) - \frac{u^2}{(uv + 2)^2} Q^2(n, v), \tag{3.19}$$

$$Q^2\left(n + 1, -\frac{1}{u}\right) = \frac{1}{u^2} Q^1(n, u). \tag{3.20}$$

Apply the operator $2(v/2 + 1/u)^2 \{u^2 \partial_u + 2 \partial_v\}$ to (3.19); this gives

$$\left(Q^1\right)'(n, u) - \frac{2}{u} Q^1(n, u) - \left(Q^2\right)'(n, v) = 0.$$

Complete the differential elimination and use (3.19) to obtain

$$Q^1(n, u) = c_1 u + A(n) u^2, \qquad Q^2(n, v) = -c_1 v + 2A(n) - A(n + 1).$$

Substitute these results into (3.20) to find the general solution of the LSC:

$$Q^1(n, u) = c_1 u + (c_2 n + c_3) u^2, \qquad Q^2(n, v) = -c_1 v + c_2(n - 1) + c_3.$$

Thus every generator of the required form is a linear combination of

$$X_1 = u\partial_u - v\partial_v, \qquad X_2 = nu^2\partial_u + (n - 1)\partial_v, \qquad X_3 = u^2\partial_u + \partial_v.$$

It turns out to be convenient to use the symmetries generated by X_3 to simplify the system[2]. This generator has canonical coordinates $(r, s) = (v + 1/u, v)$, in terms of which (3.18) amounts to

$$r_1 = r, \qquad s_1 - s = r.$$

Therefore $r = \tilde{c}_1$ and $s = \tilde{c}_2 - \tilde{c}_1 n$, and so the general solution of (3.18) is

$$u = \frac{1}{\tilde{c}_1(n + 1) - \tilde{c}_2}, \qquad v = \tilde{c}_2 - \tilde{c}_1 n. \qquad \blacktriangle$$

Canonical coordinates do not exist at any invariant point, that is, where

$$Q^\alpha(n, \mathbf{u}) = 0, \qquad \alpha = 1, \dots, M, \tag{3.21}$$

because $Xs = 1$ cannot be satisfied at such points. However, it is common for

---

[2] The reason why some generators give particularly simple reductions is discussed in §3.5.

the system (3.21) to define solutions of (3.14) that are composed entirely of invariant points. These *invariant solutions* are not always part of the general solution. (The same is true for scalar OΔEs – see Example 2.8.)

Example 3.3   The constrained system of OΔEs,

$$u_1 = \frac{u}{v-u+1}, \qquad v_1 = \frac{v}{v-u+1}, \qquad v > u, \qquad n \geq 0, \qquad (3.22)$$

has Lie symmetries generated by $X = u^2 \partial_u + uv \partial_v$. In any neighbourhood in which $u \neq 0$, the canonical coordinates $(r, s) = (v/u, -1/u)$ reduce (3.22) to

$$r_1 = r, \qquad s_1 - s = 1 - r,$$

whose general solution is $r = c_1$, $s = (1 - c_1)n + c_2$. Thus the general solution of (3.22) is

$$u = \frac{1}{(c_1 - 1)n - c_2}, \qquad v = \frac{c_1}{(c_1 - 1)n - c_2}, \qquad (c_1 - 1)c_2 < 0. \quad (3.23)$$

Every point with $u = 0$ is invariant; on the set of all such points that satisfy the constraint $v > 0$, (3.22) reduces to $v_1 = v/(v + 1)$. Consequently, every solution that is invariant under the symmetries generated by X is of the form

$$u = 0, \qquad v = \frac{1}{n+c}, \qquad c > 0.$$

Such solutions do not belong to (3.23), though they can be regarded as limiting cases of (3.23) as $c_1 \to \infty$.   ▲

3.3 Dynamical symmetries

For a given first-order OΔE (or system of OΔEs), we have needed an ansatz to obtain characteristics for Lie point symmetries. This is because the unrestricted LSC has infinitely many linearly independent solutions that one cannot find without knowing the general solution of the OΔE. By contrast, the LSC for Lie point symmetries of a scalar OΔE of order $p > 1$,

$$u_p = \omega(n, u, \dots, u_{p-1}), \qquad (3.24)$$

can generally be solved without using an ansatz, yielding finitely many linearly independent characteristics. Yet if we define $u^1 = u$, $u^2 = u_1, \dots, u^p = u_{p-1}$, the pth-order OΔE (3.24) is equivalent to the first-order system

$$u_1^\alpha = u^{\alpha+1}, \qquad \alpha = 1, \dots, p-1, \qquad u_1^p = \omega(n, u^1, \dots, u^p). \quad (3.25)$$

As (3.25) has infinitely many linearly independent characteristics, where are the missing symmetries in (3.24)?

To answer this question, it is helpful to look at the action of a given point transformation on the general solution of (3.24). There are several ways to describe the action, each of which produces some useful insights. For simplicity, we shall assume that (3.24) can be written in backward form and that the whole of $\mathbb{Z}$ is a regular domain for the OΔE. However, it is not difficult to adapt what follows to any regular domain.

The most intuitive approach is to work in the space of independent and dependent variables. Each solution, $u = f(n)$, is represented in this space by its *graph*,

$$\mathcal{G}_f = \{(n, f(n)) : n \in \mathbb{Z}\} \subset \mathbb{Z} \times \mathbb{R}. \tag{3.26}$$

A transformation of the form

$$\Psi : (n, u) \mapsto (\hat{n}, \hat{u}) = (n, \psi_n(u))$$

maps $\mathcal{G}_f$ to $\mathcal{G}_{\tilde{f}}$, where $\tilde{f}(n) = \psi_n(f(n))$. If the transformation is a symmetry, $u = \tilde{f}(n)$ is a solution of (3.24) for every solution $u = f(n)$. In particular, for Lie point symmetries, $\tilde{f}$ is locally analytic in ε. Although the tangent vector field at each n is

$$X = Q(n, u)\,\partial_u, \tag{3.27}$$

which depends only on n and u, the prolongations of X depend also on shifts of u. So it is convenient to introduce the kth *prolonged spaces* $\mathbb{Z} \times \mathbb{R}^{k+1}$, with coordinates $(n, u, u_1, \ldots, u_k)$. Each solution, $u = f(n)$, is represented in the kth prolonged space by the graph

$$\mathcal{G}_f^k = \{ (n, f(n), f(n+1), \ldots, f(n+k)) : n \in \mathbb{Z} \}.$$

A generator of a one-parameter Lie group of symmetries is then regarded as a vector field on the space $\mathbb{R}^{k+1}$ over each n, treating $u, u_1, \ldots, u_k$ as independent coordinates[3]. The generator (3.27) prolongs to

$$X = Q(n, u)\,\partial_u + Q(n+1, u_1)\,\partial_{u_1} + \cdots + Q(n+k, u_k)\,\partial_{u_k}.$$

From here on, whenever X is applied to an expression that involves shifts of u, it will be assumed to be sufficiently prolonged to act on all shifts. In particular, the LSC is determined by applying the pth prolongation of X to the OΔE, then evaluating the result on solutions of the OΔE.

---

[3] For every solution of (3.24) and every $k \geq p$, the value of u_k is determined by $u, \ldots, u_{p-1}$. Nevertheless, it is sometimes useful to regard these solutions as being embedded in a higher-dimensional space.

Given p initial values, $u(n_0), \ldots, u(n_0 + p - 1)$, the OΔE (3.24) enables us to construct all other values of u. Suppose that these initial values are mapped (by a transformation that fixes n) to p new initial values. Then we can again use (3.24) to construct a solution based on the new initial conditions. Thus any such transformation of the initial conditions corresponds to a symmetry of the OΔE. The initial conditions determine the values of the p arbitrary constants in the general solution of the OΔE, so a transformation of the initial conditions corresponds to a transformation of the arbitrary constants. We have seen that the general solution of the OΔE may be written as

$$\phi^i(n, u, \ldots, u_{p-1}) = c_i, \qquad i = 1, \ldots, p,$$

where $\phi^1, \ldots, \phi^p$ are functionally independent first integrals. So one can regard Lie symmetries as Lie point transformations in the space of first integrals. Indeed, any locally defined diffeomorphism,

$$\Gamma : \phi^i \mapsto \hat{\phi}^i = f^i(\phi^1, \ldots, \phi^p), \qquad i = 1, \ldots, p,$$

merely changes the values of the arbitrary constants, mapping $\phi^i = c_i$ to

$$\hat{\phi}^i = \hat{c}_i \equiv f^i(c_1, \ldots, c_p).$$

Let $\mathcal{C}$ denote the set of all p-tuples $\mathbf{c} = (c_1, \ldots, c_p)$ that correspond to a solution of the OΔE. Then if Γ is a symmetry, it maps $\mathcal{C}$ to itself. Lie point transformations of the first integrals are of the form

$$\hat{\phi}^i = \phi^i + \varepsilon \zeta^i(\phi^1, \ldots, \phi^p) + O(\varepsilon^2), \tag{3.28}$$

so the infinitesimal generator is

$$X = \zeta^j(\phi^1, \ldots, \phi^p)\, \partial_{\phi^j}. \tag{3.29}$$

The $(p - 1)^{\text{th}}$ prolonged space includes all of the variables in (3.29). So the change-of-variables formula,

$$X = (Xu)\, \partial_u + (Xu_1)\, \partial_{u_1} + \cdots + (Xu_{p-1})\, \partial_{u_{p-1}}, \tag{3.30}$$

yields the characteristic

$$Q = Xu = \zeta^j(\phi^1, \ldots, \phi^p)\, \frac{\partial u}{\partial \phi^j}, \tag{3.31}$$

where u is regarded as a function of $(n, \phi^1, \ldots, \phi^p)$. Similarly, the shifts of (3.31) are

$$\overline{S}_n^i Q = Xu^i = \zeta^j(\phi^1, \ldots, \phi^p)\, \frac{\partial u_i}{\partial \phi^j}, \qquad i = 1, \ldots, p - 1, \tag{3.32}$$

where $\bar{S}_n$ is the restricted shift operator. Nowhere in the derivation of (3.29) have we assumed that X generates Lie point symmetries; in general, the characteristic (3.31) depends (at most) on $n, u, u_1, \ldots, u_{p-1}$. The symmetries generated by X are called *dynamical symmetries*[4]; they constitute a one-parameter local Lie group that acts on the $(p-1)^{\text{th}}$ prolonged space. Just as for first-order OΔEs, it is fairly easy to find all dynamical symmetry characteristics if the first integrals are known.

Example 3.4 The OΔE

$$u_2 - 2u_1 + u = 0$$

has the first integrals

$$\phi^1 = u_1 - u, \qquad \phi^2 = (n+1)u - nu_1.$$

Thus $u = \phi^1 n + \phi^2$, so the most general characteristic is

$$Q(n, u, u_1) = n\zeta^1(u_1 - u, (n+1)u - nu_1) + \zeta^2(u_1 - u, (n+1)u - nu_1), \quad (3.33)$$

where ζ^1, ζ^2 are arbitrary smooth functions. Characteristics of a particular form can be found from (3.33). For instance, the characteristics that are linear in u_1 are found by differentiating (3.33) twice with respect to u_1; this yields

$$n^3\zeta^1_{,22} + n^2(\zeta^2_{,22} - 2\zeta^1_{,12}) + n(\zeta^1_{,11} - 2\zeta^2_{,12}) + \zeta^2_{,11} = 0. \quad (3.34)$$

The functions ζ^j are independent of n, so (3.34) splits into the system

$$\zeta^1_{,22} = 0, \quad \zeta^2_{,22} - 2\zeta^1_{,12} = 0, \quad \zeta^1_{,11} - 2\zeta^2_{,12} = 0, \quad \zeta^2_{,11} = 0,$$

whose general solution is

$$\zeta^1(\phi^1, \phi^2) = c_1 + c_3\phi^1 + c_6\phi^2 + c_5(\phi^1)^2 + c_8\phi^1\phi^2,$$
$$\zeta^2(\phi^1, \phi^2) = c_2 + c_7\phi^2 + c_4\phi^1 + c_5\phi^1\phi^2 + c_8(\phi^2)^2.$$

In terms of (n, u, u_1), the characteristic (3.33) is

$$Q = c_1 n + c_2 + (c_3 n + c_4 + c_5 u)(u_1 - u) + (c_6 n + c_7 + c_8 u)((n+1)u - nu_1). \quad \blacktriangle$$

The same approach is used to prove some general results about dynamical symmetries of various types of OΔEs. For example, for *every* second-order linear OΔE of the form

$$u_2 + a_1(n)u_1 + a_2(n)u = 0, \qquad a_1(n) \neq 0, \ a_2(n) \neq 0, \quad (3.35)$$

the most general characteristic that is linear in u_1 depends upon eight arbitrary

---

[4] A point symmetry is a type of dynamical symmetry, because it acts as a point transformation on the space of first integrals.

constants. If $u = f_1(n)$ and $u = f_2(n)$ are two linearly independent solutions of (3.35), the characteristic is

$$Q = c_1 f_1(n) + c_2 f_2(n) + \{c_3 f_1(n) + c_4 f_2(n) + c_5 u\} \, \phi^1(n, u, u_1)$$
$$+ \{c_6 f_1(n) + c_7 f_2(n) + c_8 u\} \, \phi^2(n, u, u_1), \tag{3.36}$$

where

$$\phi^1(n, u, u_1) = \frac{f_2(n+1)\, u - f_2(n)\, u_1}{f_1(n) f_2(n+1) - f_1(n+1) f_2(n)},$$
$$\phi^2(n, u, u_1) = \frac{f_1(n)\, u_1 - f_1(n+1)\, u}{f_1(n) f_2(n+1) - f_1(n+1) f_2(n)}.$$

The proof of this result is left as an exercise.

Generally speaking, the first integrals are not known, so one must seek dynamical symmetries by solving the LSC,

$$\overline{S}_n^p Q = \sum_{k=0}^{p-1} \omega_{,k+1} \overline{S}_n^k Q. \tag{3.37}$$

Again, the chief difficulty is to pick an appropriate ansatz for Q. If $p \geq 3$ and the OΔE is not too complicated, one might seek characteristics that are independent of u_{p-1}. When $p = 2$, it may be fairly easy to find all characteristics that are linear in u_{p-1}. However, to reduce the risk of making basic errors in the calculation, use computer algebra!

3.4 The commutator

We have looked at the action of symmetries on solutions of a given scalar OΔE from various geometrical perspectives. However, Lie groups of symmetries have an intrinsic structure that must also be taken into account. For simplicity, we shall restrict attention to Lie point symmetries at present. The set of all characteristics $Q(n, u)$ is necessarily a vector space, because the LSC is linear homogeneous in Q. When this set is R-dimensional, every characteristic can be written in the form

$$Q(n, u) = \sum_{i=1}^{R} c_i Q_i(n, u), \tag{3.38}$$

where $Q_1, \ldots, Q_R$ are linearly independent characteristics. The corresponding set of infinitesimal generators is the R-dimensional vector space $\mathcal{L}$ spanned by

$$X_i = Q_i(n, u)\, \partial_u, \qquad i = 1, \ldots, R. \tag{3.39}$$

The set G of all point symmetries that belong to a one-parameter local Lie group is obtained by exponentiating the elements of $\mathcal{L}$. Each such element is determined by R arbitrary constants, so G is said to be an *R-parameter local Lie group*.

Two elements of a group, Γ_i and Γ_j, *commute* if $\Gamma_i \Gamma_j = \Gamma_j \Gamma_i$, that is,

$$\Gamma_i^{-1} \Gamma_j^{-1} \Gamma_i \Gamma_j = \text{id.} \tag{3.40}$$

If the group elements are Lie point transformations

$$\Gamma_i = e^{\varepsilon_i X_i}, \qquad \Gamma_j = e^{\varepsilon_j X_j},$$

we can expand in powers of ε_i and ε_j to obtain

$$\Gamma_i^{-1} \Gamma_j^{-1} \Gamma_i \Gamma_j = e^{-\varepsilon_i X_i} e^{-\varepsilon_j X_j} e^{\varepsilon_i X_i} e^{\varepsilon_j X_j}$$

$$= \text{id} + \varepsilon_i \varepsilon_j (X_i X_j - X_j X_i) + \text{higher-order terms.}$$

The operator

$$[X_i, X_j] \equiv X_i X_j - X_j X_i \tag{3.41}$$

is called the *commutator* of X_i and X_j. Clearly, if the commutator is nonzero then Γ_i and Γ_j do not commute; the converse is also true[5]. It is straightforward to show that the set $\mathcal{L}$ is closed under the commutator:

$$X_i, X_j \in \mathcal{L} \implies [X_i, X_j] \in \mathcal{L}. \tag{3.42}$$

Example 3.5 For the third-order OΔE

$$u_3 = u_2 + \frac{(u_2 - u_1)^2}{u_1 - u},$$

the general solution of the LSC for Lie point symmetries is

$$Q(n, u) = c_1 + c_2 u.$$

Therefore $\mathcal{L}$ is spanned by the infinitesimal generators

$$X_1 = \partial_u, \qquad X_2 = u \partial_u.$$

The commutator of X_1 and X_2 is

$$[X_1, X_2] = \partial_u = X_1,$$

so

$$\Gamma_1 = e^{\varepsilon_1 X_1}, \qquad \Gamma_2 = e^{\varepsilon_2 X_2}$$

---

[5] The proof of this is left as an exercise.

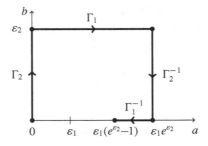

Figure 3.3 The non-commuting symmetries in Example 3.5, shown in parameter space: here a is the translation and b is the log of the scaling factor.

do not commute. Indeed,

$$\Gamma_1^{-1}\Gamma_2^{-1}\Gamma_1\Gamma_2(u) = \Gamma_1^{-1}\Gamma_2^{-1}(e^{\varepsilon_2}u + \varepsilon_1 e^{\varepsilon_2}) = u + \varepsilon_1(e^{\varepsilon_2} - 1),$$

and so

$$\Gamma_1^{-1}\Gamma_2^{-1}\Gamma_1\Gamma_2 = \exp\{\varepsilon_1(e^{\varepsilon_2} - 1)X_1\}.$$

This is illustrated in Figure 3.3, which shows that the extent of the translation Γ_1 is determined by the amount of stretch Γ_2. After each symmetry is applied, u is mapped to $e^b u + a$; the points (a, b) are plotted. As the path formed by $\Gamma_1^{-1}\Gamma_2^{-1}\Gamma_1\Gamma_2$ does not return to the origin, Γ_1 and Γ_2 do not commute.   ▲

Given a basis $X_1, \ldots, X_R$ of $\mathcal{L}$, the closure condition (3.42) implies that there exist constants c_{ij}^k (called *structure constants*) such that

$$[X_i, X_j] = c_{ij}^k X_k, \qquad 1 \le i, j \le R. \tag{3.43}$$

By definition, the commutator is *antisymmetric*, which means that

$$[X_i, X_j] = -[X_j, X_i], \tag{3.44}$$

and it satisfies the *Jacobi identity*:

$$\left[X_i, [X_j, X_k]\right] + \left[X_j, [X_k, X_i]\right] + \left[X_k, [X_i, X_j]\right] = 0. \tag{3.45}$$

Consequently the structure constants are subject to the constraints

$$c_{ij}^k = -c_{ji}^k, \tag{3.46}$$

$$c_{ij}^q c_{kq}^l + c_{jk}^q c_{iq}^l + c_{ki}^q c_{jq}^l = 0. \tag{3.47}$$

The commutator is a product on $\mathcal{L}$ that is bilinear (linear in each argument), antisymmetric, and satisfies the Jacobi identity. Any vector space that is equipped with such a product is called a *Lie algebra*. A Lie group of transformations may be regarded abstractly (in terms of its generators and relations) or concretely (in terms of the action of its elements on an object). Similarly, the associated Lie algebra is determined abstractly by its structure constants, but its realization is determined by the explicit form of the infinitesimal generators. A particular abstract Lie algebra can be realized in many different ways.

Although we have focused on Lie point symmetries, the commutator arises in the same way for all R-parameter local Lie groups of point transformations. Dynamical symmetries are point transformations in the space of first integrals, so the commutator of any two generators of dynamical symmetries is itself a dynamical symmetry generator. The Lie algebra of all such generators is infinite-dimensional, because the general solution of the LSC depends on p arbitrary functions. However, the set of dynamical symmetry generators of a particular form may be an R-dimensional Lie algebra.

Example 3.6 For the OΔE

$$u_2 = 2u_1^3/u^2 - u_1,$$

the set of all characteristics that are linear in u_1 is spanned by

$$Q_1 = u_1/u, \qquad Q_2 = u.$$

The corresponding generators, prolonged to include their action on u_1 as well as u, are

$$X_1 = (u_1/u)\,\partial_u + \left(2u_1^2/u^2 - 1\right)\partial_{u_1}, \qquad X_2 = u\partial_u + u_1\partial_{u_1}.$$

A short calculation yields

$$[X_1, X_2] = X_1,$$

so in this case, the set of all generators whose characteristics are linear in u_1 is a Lie algebra with the same structure constants as the one in Example 3.5. ▲

Note: The commutator of two dynamical symmetries that are linear in u_1 is a dynamical symmetry that may or may not be linear in u_1. More generally, the set of all dynamical symmetries satisfying a particular ansatz need not be closed under the commutator.

Groups that are not simple can be described in terms of their subgroups; normal subgroups are of particular importance. Similar ideas apply to Lie algebras; they are the key to carrying out multiple reductions of order. A *Lie subalgebra* of a given Lie algebra $\mathcal{L}$ is a subspace, $\mathcal{M} \subset \mathcal{L}$, that is a Lie algebra in its own right. Hence

$$[\mathcal{M}, \mathcal{M}] \subset \mathcal{M}, \tag{3.48}$$

where $[\mathcal{A}, \mathcal{B}]$ denotes the vector space spanned by the commutators of all elements of $\mathcal{A}$ with all elements of $\mathcal{B}$. For instance, if all structure constants are zero (that is, if $\mathcal{L}$ is abelian), every subspace of $\mathcal{L}$ is an abelian Lie subalgebra.

An *ideal* is a particularly important type of Lie subalgebra, with the stronger property that

$$[\mathcal{M}, \mathcal{L}] \subset \mathcal{M}. \qquad (3.49)$$

(Ideals are the counterparts of normal subgroups.) Trivially, $\mathcal{L}$ and $\{0\}$ are ideals of $\mathcal{L}$. If a non-abelian Lie algebra has no nontrivial ideals, it is said to be *simple*. Given $\mathcal{L}$, one can always construct the series of ideals

$$\mathcal{L}^{(0)} = \mathcal{L}, \qquad \mathcal{L}^{(i+1)} = [\mathcal{L}^{(i)}, \mathcal{L}^{(i)}],$$

stopping when $\mathcal{L}^{(i+1)} = \mathcal{L}^{(i)}$. This is called the *derived series*. Any Lie algebra for which the derived series stops at $\{0\}$ is said to be *solvable*. If $\mathcal{L}$ is solvable, one can use the derived series to construct a chain of subalgebras,

$$\{0\} = \mathcal{L}_0 \subset \mathcal{L}_1 \subset \cdots \subset \mathcal{L}_R = \mathcal{L}, \qquad (3.50)$$

such that each $\mathcal{L}_k$ is a k-dimensional ideal of $\mathcal{L}_{k+1}$. Given such a chain, a *canonical basis* is a basis for $\mathcal{L}$ such that $\mathcal{L}_k = \mathrm{Span}(X_1, \ldots, X_k)$, for $k = 1, \ldots, R$.

Example 3.7   The Lie algebra $\mathcal{L}$ of point symmetry generators for

$$u_5 = u_3 + \frac{(u_4 - u_2)(u_3 - u_1)}{u_2 - u} \qquad (3.51)$$

is spanned by

$$X_1 = \partial_u, \quad X_2 = (-1)^n \partial_u, \quad X_3 = u\partial_u, \quad X_4 = (-1)^n u\partial_u. \qquad (3.52)$$

For this Lie algebra, the only nonzero structure constants c_{ij}^k (with $i < j$) are

$$c_{13}^1 = c_{14}^2 = c_{23}^2 = c_{24}^1 = 1.$$

Therefore the derived series has

$$\mathcal{L}^{(0)} = \mathcal{L}, \qquad \mathcal{L}^{(1)} = \mathrm{Span}(\partial_u, (-1)^n \partial_u), \qquad \mathcal{L}^{(2)} = \{0\}.$$

To produce a chain of ideals (3.50), start with the observation that $\mathcal{L}_0 = \mathcal{L}^{(2)}$. Now pick an element of the previous ideal in the derived series (that is, $\mathcal{L}^{(1)}$) to form a basis for $\mathcal{L}_1$. Continue in this way until $\mathcal{L}^{(1)}$ is exhausted; one possibility is to take

$$\mathcal{L}_1 = \mathrm{Span}(\partial_u), \quad \mathcal{L}_2 = \mathrm{Span}(\partial_u, (-1)^n \partial_u) = \mathcal{L}^{(1)}.$$

Now continue the process, working up the derived series until $\mathcal{L}_4 = \mathcal{L}$ is obtained. One possibility is

$$\mathcal{L}_3 = \text{Span}(\partial_u, (-1)^n \partial_u, u\partial_u), \quad \mathcal{L}_4 = \mathcal{L}.$$

This shows that the basis (3.52) is a canonical basis. It is not the only canonical basis, because we could have chosen another element of $\mathcal{L}^{(1)}$ as a basis for $\mathcal{L}_1$; a similar choice was faced in the construction of $\mathcal{L}_3$. ▲

3.5 Multiple reductions of order

In Chapter 2, we saw that any one-parameter Lie group of point symmetries can be used to reduce the order of the OΔE once. If we know two symmetry generators, can we reduce the order twice? It turns out that the answer to this question is determined by the structure of the Lie algebra. Before considering the general case, it is helpful to look at an example.

Example 3.8 In Example 3.5, it was shown that the Lie algebra of point symmetry generators for

$$u_3 = u_2 + \frac{(u_2 - u_1)^2}{u_1 - u} \tag{3.53}$$

is spanned by

$$X_1 = \partial_u, \qquad X_2 = u\partial_u.$$

Let us reduce the order once by using X_1, whose first prolongation has canonical coordinates

$$r = u_1 - u, \qquad s = u.$$

The OΔE amounts to

$$r_2 = \frac{r_1^2}{r}. \tag{3.54}$$

We have already seen how to solve this reduced OΔE, so r can be found and the original OΔE can be solved completely. However, one reason why the reduced OΔE is tractable is that it inherits the symmetries generated by X_2. To see this, apply X_2 (prolonged to include the action on u_1) to r:

$$X_2 r = u \frac{\partial r}{\partial u} + u_1 \frac{\partial r}{\partial u_1} = u_1 - u = r.$$

Therefore X_2 reappears as the scaling symmetry generator

$$\tilde{X}_2 = r\partial_r,$$

which enables the reduced equation (3.54) to be reduced in order a second time.

By contrast, let us use canonical coordinates for the first prolongation of X_2, namely

$$\tilde{r} = \frac{u_1}{u}, \qquad \tilde{s} = \ln |u|,$$

to reduce (3.53). This reduction of order works (as it is guaranteed to do), yielding

$$\tilde{r}_2 = 1 + \frac{\tilde{r}(\tilde{r}_1 - 1)^2}{\tilde{r}_1(\tilde{r} - 1)}. \qquad (3.55)$$

Unlike (3.54), the reduced OΔE (3.55) has no Lie point symmetries. What has happened to the symmetries generated by X_1? As before, apply X_1 (prolonged) to $\tilde{r}$, to get

$$X_1\tilde{r} = \frac{\partial \tilde{r}}{\partial u} + \frac{\partial \tilde{r}}{\partial u_1} = \frac{1}{u} - \frac{u_1}{u^2} = \frac{1 - \tilde{r}}{u}.$$

This is not a function of $\tilde{r}$ only, so it cannot be a generator of Lie point symmetries for the reduced OΔE (3.55). In summary, if one writes the OΔE in terms of the invariant r of X_1, the symmetries generated by X_2 are inherited; the symmetries generated by X_1 are not inherited if X_2 is used first.   ▲

The above situation occurs for any Lie algebra $\mathcal{L} = \text{Span}(X_1, X_2)$ for which

$$[X_1, X_2] = X_1. \qquad (3.56)$$

However, if the two generators commute, that is, if

$$[X_1, X_2] = 0, \qquad (3.57)$$

the reduction can be carried out in either order; the symmetries corresponding to the unused generator will be inherited by the reduced equation[6].

For any two-dimensional Lie algebra, $\mathcal{L}$, it is possible to choose a basis such that either (3.56) or (3.57) holds. (The proof of this result is left as an exercise.) So if $\dim(\mathcal{L}) = 2$, the Lie algebra is solvable and two reductions of order are always possible. Indeed, further reductions may become apparent if one of the reduced OΔEs has extra point symmetries.

The same idea carries over to higher-dimensional solvable Lie algebras. Given a canonical basis, write the given p^{th}-order OΔE in terms of canonical coordinates, $(r^1(n, u, u_1),\ s^1(n, u))$, that are derived from the prolonged X_1. This produces a reduced OΔE which inherits all of the remaining symmetry generators. Then use canonical coordinates for the prolonged X_2 that has been

---

[6] This does not imply that both reductions will be equally easy to use – see Hydon (2000b).

written in terms of r^1, namely $(r^2(n, r^1, r_1^1)$, $s^2(n, r^1))$, to generate a second reduction of order, and so on. Provided that the solvable Lie algebra is of dimension p or more, this process yields the solution of the original OΔE.

Some useful shortcuts are available. In particular, it is often convenient to write the original OΔE in terms of a function $v(n, u, u_1, \ldots, u_k)$ that is invariant under $X_1, \ldots, X_k$ (suitably prolonged), without going one step at a time. For instance, $v = u_2 - u$ is invariant under ∂_u and $(-1)^n \partial_u$. Furthermore, if a reduced OΔE has an obvious general solution, one can terminate the process early.

Not every OΔE has a solvable Lie algebra of Lie point symmetry generators (see Exercise 3.14 for an OΔE with a simple Lie algebra). Sometimes, however, one can use a solvable subalgebra to reduce the OΔE to something manageable.

3.6 Partitioned OΔEs

So far, the theory of Lie symmetries for OΔEs corresponds to the theory for ODEs. This section deals with an exception, which requires the independent variable to be discrete. A forward OΔE is *partitioned* if it is of the form

$$u_{qL} = \omega(n, u, u_L, u_{2L}, \ldots, u_{(q-1)L}), \qquad n \in \mathbb{Z}, \tag{3.58}$$

for some $L \geq 2$. In other words, the OΔE involves only values of u at points that are separated by a multiple of L. Such an OΔE can be rewritten as a set of L scalar OΔEs of order q, as follows. Define L dependent variables, $u^{(0)}, u^{(1)}, \ldots, u^{(L-1)}$, as follows:

$$u^{(l)}(m) = u(mL + l), \qquad m \in \mathbb{Z}.$$

Then for $n = mL + l$, the OΔE (3.58) amounts to

$$u_q^{(l)} = \omega(mL + l, u^{(l)}, u_1^{(l)}, u_2^{(l)}, \ldots, u_{q-1}^{(l)}), \qquad n \in \mathbb{Z}. \tag{3.59}$$

Example 3.9 The simplest example of a partitioned OΔE is

$$u_2 - u = 0. \tag{3.60}$$

The above transformation, with $L = 2$, reduces (3.60) to the system

$$u_1^{(l)} - u^{(l)} = 0, \qquad l = 0, 1. \tag{3.61}$$

Consequently, the general solution of (3.60) is

$$u = \begin{cases} c_1, & n \text{ even}, \\ c_2, & n \text{ odd}. \end{cases}$$

It is helpful to rewrite this as a single function,

$$u = c_1 \pi_E(n) + c_2 \pi_O(n),$$

where

$$\pi_E(n) = (1 + (-1)^n)/2, \qquad \pi_O(n) = (1 - (-1)^n)/2. \qquad (3.62)$$

We call π_E the *even projector* and π_O the *odd projector*; these functions satisfy

$$(\pi_E(n))^2 = \pi_E(n), \qquad (\pi_O(n))^2 = \pi_O(n),$$
$$\pi_E(n)\pi_O(n) = 0, \qquad \pi_E(n) + \pi_O(n) = 1. \qquad (3.63)$$

They are particularly useful for demonstrating some of the contrasts between differential equations and difference equations.

From our earlier results on symmetries of linear OΔEs, it is natural to suppose that the general solution of the LSC for point symmetries of (3.60) is

$$Q(n, u) = c_1 \pi_E(n) + c_2 \pi_O(n) + c_3 u. \qquad (3.64)$$

However, although (3.64) satisfies the LSC, it is not the general solution, because the most general characteristic for each of the OΔEs in (3.61) is

$$Q(m, u^{(l)}) = f_l(u^{(l)}), \qquad (3.65)$$

where f_l is an arbitrary smooth function. Therefore, in terms of the original variables, the general solution of the LSC is

$$Q(n, u) = \pi_E(n)f_0(u) + \pi_O(n)f_1(u). \qquad (3.66)$$

Clearly, (3.64) is only a small subset of these solutions.　　　　　　▲

Partitioning produces something of a paradox: there are qL arbitrary constants in the general solution of a $(qL)^{\text{th}}$-order partitioned OΔE, yet the dimension of the Lie algebra is determined in part by the partitioning.

Example 3.10 To illustrate this, consider Lie point symmetries of the OΔEs

$$u_6 - 2u_k + u = 0, \qquad 1 \leq k \leq 3. \qquad (3.67)$$

When $k = 1$, the OΔE (3.67) is not partitioned and so $\dim(\mathcal{L}) = 7$. For $k = 2$, (3.67) is reducible to a pair of third-order OΔEs, each of which has a four-dimensional Lie algebra; hence $\dim(\mathcal{L}) = 8$. When $k = 3$, (3.67) reduces to three second-order OΔEs, each with a three-dimensional Lie algebra, and so $\dim(\mathcal{L}) = 9$.　　　　　　▲

Where a partitioned OΔE (3.58) does not involve n explicitly, each of the qth-order equations (3.59) is identical, so the Lie algebras also coincide. Thus, knowledge of one characteristic is enough to reduce each of (3.59) by one order, which is equivalent to reducing the original OΔE by L orders. This does not happen if (3.58) depends on n, as the following example shows.

Example 3.11   The partitioned OΔE

$$u_4 = \left(2 + \frac{\pi_0(n)}{u}\right) u_2 - u \tag{3.68}$$

splits into

$$u_2^{(0)} = 2u_1^{(0)} - u^{(0)}, \qquad u_2^{(1)} = \left(2 + \frac{1}{u^{(1)}}\right) u_1^{(1)} - u^{(1)}.$$

The second of these has no Lie point symmetries, so every characteristic of Lie point symmetries for (3.68) is of the form

$$Q(n, u) = (c_1 n + c_2 + c_3 u) \pi_E(n).$$

These can be used to deduce the form of the solution for all even n, namely $u = \tilde{c}_1 n + \tilde{c}_2$, but the solution for odd n remains unknown.   ▲

Thus, in effect, partitioning splits the problem of solving a given OΔE into disjoint pieces. It may be possible to solve some pieces, but not others.

3.7 Variational OΔEs

For ODEs, Noether's Theorem links symmetries of a given variational problem with first integrals of the Euler–Lagrange equation(s). There is a corresponding result for OΔEs. This is a special instance of Noether's Theorem for PΔEs, which is described in detail (with proofs of the main results) in Chapter 6. However, the key ideas for scalar OΔEs are easily understood at this stage, provided that we defer some proofs until later.

The simplest type of difference variational problem is as follows. Given a real-valued function, $L(n, u, u_1)$, which is smooth in u and u_1 (for each n), find all extrema of the functional

$$\mathcal{L}[u] = \sum_{n=n_0}^{n_1} L(n, u, u_1). \tag{3.69}$$

Here, n_0 (resp. n_1) may be replaced by $-\infty$ (resp. ∞), provided that the sum converges for every u that is sufficiently close to an extremum. The term 'close to' is defined in terms of the norm that is induced by the ℓ^2 inner product

$$\langle u, v \rangle = \sum_{n=n_0}^{n_1+1} u(n)\, v(n), \tag{3.70}$$

namely $\| u \| = (\langle u, u \rangle)^{1/2}$. For simplicity, we use formal arguments henceforth, assuming that questions of convergence have already been addressed.

Extrema can occur only where the functional $\mathcal{L}[u]$ is stationary with respect to all sufficiently small allowable variations. This means that

$$\left\{ \frac{\mathrm{d}}{\mathrm{d}\epsilon}\, \mathcal{L}[u + \epsilon \eta] \right\}\bigg|_{\epsilon=0} = 0, \tag{3.71}$$

for all functions $\eta(n)$ such that $u + \epsilon \eta$ satisfies any boundary conditions that are imposed on u. Consequently, for every such η,

$$\sum_{n=n_0}^{n_1} \{ L_{,1}\, \eta + L_{,2}\, \eta_1 \} = 0. \tag{3.72}$$

By changing the variable over which the second term is summed, (3.72) may be rewritten as

$$\sum_{n=n_0+1}^{n_1} \left\{ L_{,1} + S_n^{-1}(L_{,2}) \right\} \eta + (L_{,1})|_{n=n_0} \eta(n_0) + (L_{,2})|_{n=n_1} \eta(n_1+1) = 0. \tag{3.73}$$

The function η can take arbitrary values at each point, except possibly the boundary points n_0 and $n_1 + 1$. So (3.73) will hold for all allowable η only if

$$L_{,1} + S_n^{-1}(L_{,2}) = 0, \qquad n_0 + 1 \leq n \leq n_1. \tag{3.74}$$

This second-order OΔE is called the *Euler–Lagrange equation*; L is called the *Lagrangian*. If u is unconstrained at the boundaries, the remaining terms in (3.73) vanish if and only if the following *natural boundary conditions* hold:

$$(L_{,1})|_{n=n_0} = 0, \qquad (L_{,2})|_{n=n_1} = 0. \tag{3.75}$$

(When n_0 (resp. n_1) is replaced by $-\infty$ (resp. ∞), the corresponding limiting natural boundary condition is used whenever the limit of u is not prescribed.)

Commonly, η is constrained by $u(n_0)$ and $u(n_1 + 1)$ being specified. In this circumstance, η is allowable only if

$$\eta(n_0) = 0, \qquad \eta(n_1 + 1) = 0. \tag{3.76}$$

Prescribed boundary conditions for $u(n_0)$ and $u(n_1 + 1)$ are called *essential boundary conditions*.

Example 3.12 Find the extrema of (3.69) for each of the Lagrangians

(i) $L = uu_1 - u^2$,
(ii) $L = \exp\{u - u_1\}$,

with $n_0 = 0$, $n_1 = 1000$, subject to the essential boundary conditions $u(0) = 0$, $u(1001) = 1001$. What happens if no boundary conditions are prescribed?

Solution: In case (i), the Euler–Lagrange equation is

$$L_{,1} + S_n^{-1}(L_{,2}) \equiv u_1 - 2u + u_{-1} = 0, \qquad 1 \le n \le 1000.$$

The general solution of this OΔE is

$$u = c_1 n + c_2,$$

so the solution that satisfies the prescribed boundary conditions is

$$u = n.$$

In case (ii), the Euler–Lagrange equation is

$$L_{,1} + S_n^{-1}(L_{,2}) \equiv e^{u - u_1} - e^{u_{-1} - u} = 0, \qquad 1 \le n \le 1000,$$

which (for real u) is equivalent to the equation in case (i); hence, the solution of the boundary-value problem is the same. This illustrates an interesting feature of variational calculus: different Lagrangians can produce equivalent Euler–Lagrange equations.

If no boundary conditions are prescribed, the natural boundary conditions for the Lagrangian in (i) are

$$u(1) - 2u(0) = 0, \qquad u(1000) = 0,$$

so the extremal solution is $u = 0$. For the Lagrangian in (ii), however, the natural boundary conditions,

$$e^{u(0) - u(1)} = 0, \qquad -e^{u(1000) - u(1001)} = 0,$$

cannot be satisfied, so no extremal solution exists for the unconstrained variational problem. ▲

The above example shows the role of essential boundary conditions in picking out solutions that are extrema of the functional $\mathscr{L}[u]$. It is convenient to study all such extrema at the same time, which we will do by imposing the conditions (3.76) without explicitly stating the values of $u(n_0)$ and $u(n_1 + 1)$. Then the extrema are solutions of the Euler–Lagrange equation (without any natural boundary conditions).

Note: A less restrictive approach would be to use *Lagrange multipliers* to apply all essential boundary conditions. To illustrate this approach, let us reconsider Example 3.12(i), with the modified functional

$$\widetilde{\mathscr{L}}[u, \lambda, \mu] = \sum_{n=0}^{1000} \{uu_1 - u^2 - \lambda u(0) - \mu\{u(1001) - 1001\}\},$$

where λ and μ are functions of n. Taking arbitrary variations with respect to u gives

$$u_1 - 2u + u_{-1} = 0, \qquad 1 \leq n \leq 1000, \qquad (3.77)$$

$$u(1) - 2u(0) - \lambda(0) = 0, \qquad u(1000) - \mu(1000) = 0. \qquad (3.78)$$

Arbitrary variations of the Lagrange multipliers λ and μ enforce the essential boundary conditions,

$$u(0) = 0, \qquad u(1001) - 1001 = 0. \qquad (3.79)$$

Lagrange multipliers enable us to incorporate essential and natural boundary conditions within the same framework. When an essential boundary condition is given, the corresponding Lagrange multiplier cancels out the terms arising from the natural boundary condition, as shown by (3.78).

In Chapter 6, we will use Lagrange multipliers to enforce other constraints. However, it is most convenient to apply essential boundary conditions directly by restricting η according to (3.76). For simplicity, the limits of sums will not usually be stated from here on; instead, we write

$$\mathscr{L}[u] = \sum L(n, u, u_1)$$

and proceed formally, with the assumption that there are no contributions from boundary terms.

Just as in the continuous case, some Lagrangians do not contribute to the Euler–Lagrange equation. For instance, if

$$L(n, u, u_1) = (S_n - I)f(n, u) = f(n + 1, u_1) - f(n, u), \qquad (3.80)$$

then

$$L_{,1} + S_n^{-1}(L_{,2}) \equiv 0. \qquad (3.81)$$

It is a simple exercise to show that *every* Lagrangian which satisfies (3.81) for all n (from n_0 to n_1) is of the form (3.80). Such a Lagrangian is called a *null Lagrangian*. The freedom to add null Lagrangians to a given Lagrangian enables us to put L into a standard form in which each term involves u, as the following example illustrates.

Example 3.13 The Lagrangian

$$\tilde{L} = uu_1 - \alpha \ln\left((u_1)^2 + 1\right), \qquad \alpha \neq 0,$$

yields the Euler–Lagrange equation

$$u_1 + u_{-1} - \frac{2\alpha u}{u^2 + 1} = 0, \tag{3.82}$$

which is a McMillan map (see Exercise 2.16). The standard form of the Lagrangian is

$$L = uu_1 - \alpha \ln\left(u^2 + 1\right), \tag{3.83}$$

which differs from $\tilde{L}$ by a null Lagrangian. ▲

Everything that we have seen for first-order Lagrangians generalizes immediately to Lagrangians of higher order. The functional

$$\mathscr{L}[u] = \sum L(n, u, ..., u_p) \tag{3.84}$$

is stationary with respect to sufficiently small variations $\epsilon\eta$ if

$$\sum \sum_{j=0}^{p} \frac{\partial L}{\partial u_j} \eta_j = 0. \tag{3.85}$$

As the function η is arbitrary (except at the boundaries), we obtain the Euler–Lagrange equation

$$\mathbf{E}_u(L) \equiv \sum_{j=0}^{p} \mathrm{S}_n^{-j} \frac{\partial L}{\partial u_j} = 0. \tag{3.86}$$

The linear operator $\mathbf{E}_u$ is called the *Euler operator*. In Chapter 6, we prove that $\mathbf{E}_u(L) \equiv 0$ (that is, the Lagrangian is null) if and only if there exists f such that

$$L = (\mathrm{S}_n - \mathrm{I})f(n, u, ..., u_{p-1}). \tag{3.87}$$

3.8 Variational symmetries and first integrals

At first sight, it might appear that any symmetry of the Euler–Lagrange equation will be a symmetry of the variational problem. However, it is possible for a symmetry to map the Euler–Lagrange equation to a different (but equivalent) equation. If this occurs, either the equivalent equation is not an Euler–Lagrange equation, or else its Lagrangian differs from the original one. In either case, the original variational problem is changed significantly.

A transformation that maps the Euler–Lagrange equation to itself (not to another equation that has the same solutions) is called a *variational symmetry*.

If X generates variational symmetries, the functionals $\mathcal{L}[u] = \sum L$ and

$$\mathcal{L}[\hat{u}] = \sum \left\{ L + \varepsilon XL + O(\varepsilon^2) \right\} \qquad (3.88)$$

produce the same Euler–Lagrange equation, for each ε sufficiently close to zero. Consequently,

$$\mathbf{E}_u(XL) \equiv 0. \qquad (3.89)$$

Thus, there exists a function f such that

$$XL = (S_n - I)f. \qquad (3.90)$$

It is easy to show that if (3.89) holds, the higher-order terms in (3.88) lie in the kernel of the Euler operator. Consequently, (3.89) is both necessary and sufficient for X to generate variational symmetries, and $\mathcal{L}[\hat{u}] - \mathcal{L}[u]$ involves only boundary terms. Furthermore, if (3.89) holds, X generates symmetries of the Euler–Lagrange equation, so every variational symmetry is an ordinary symmetry. The proof of these results is left as an exercise.

To find variational symmetries, first seek solutions of the linearized symmetry condition for the Euler–Lagrange equation, then check to see which of these satisfy (3.89). It is not necessary to restrict attention to point symmetries; everything that we have derived so far is independent of the precise form of X.

Example 3.14 For $L = uu_1 - (u)^2$, the Euler–Lagrange equation is

$$u_1 - 2u + u_{-1} = 0. \qquad (3.91)$$

Every point symmetry characteristic for (3.91) is a linear combination of

$$Q_1 = 1, \quad Q_2 = n, \quad Q_3 = u.$$

The first two of these produce variational symmetries, because

$$X_1 L = u_1 - u = (S_n - I)u, \qquad (3.92)$$

$$X_2 L = nu_1 - (n - 1)u = (S_n - I)((n - 1)u). \qquad (3.93)$$

However, X_3 does not generate variational symmetries:

$$X_3 L = 2uu_1 - 2u^2 = 2L$$

is not a null Lagrangian. (Note that $\mathbf{E}_u(X_3 L) = 2\mathbf{E}_u(L)$ vanishes only on solutions of (3.91).) The group generated by X_3 rescales the Lagrangian, so the Euler–Lagrange equation is also rescaled. As an example of a non-point variational symmetry, consider the characteristic $Q_4 = u_1 - u_{-1}$, which gives

$$X_4 L = (S_n - I)(u_{-1}u_1 - 2u_{-1}u + u^2). \qquad \blacktriangle$$

At this stage, we have enough information to use variational symmetries to construct first integrals. Starting with (3.90), which amounts to

$$\sum_{j\geq 0}\left(S_n^j Q\right)\frac{\partial L}{\partial u_j} = (S_n - I)f, \tag{3.94}$$

one can use the summation by parts formula,

$$(S\,A)B = (S_n - I)\left\{A\left(S_n^{-1}B\right)\right\} + A\left(S_n^{-1}B\right), \tag{3.95}$$

to obtain, for each $j \geq 1$,

$$\left(S_n^j Q\right)\frac{\partial L}{\partial u_j} = (S_n - I)\left\{\sum_{i=0}^{j-1}\left(S_n^i Q\right)\left(S_n^{i-j}\,\frac{\partial L}{\partial u_j}\right)\right\} + Q\left(S_n^{-j}\,\frac{\partial L}{\partial u_j}\right). \tag{3.96}$$

Therefore (3.94) yields

$$Q\,E_u(L) \equiv \sum_{j\geq 0} Q\left(S_n^{-j}\,\frac{\partial L}{\partial u_j}\right) = (S_n - I)\phi, \tag{3.97}$$

where

$$\phi = f - \sum_{j\geq 1}\sum_{i=0}^{j-1}\left(S_n^i Q\right)\left(S_n^{i-j}\,\frac{\partial L}{\partial u_j}\right). \tag{3.98}$$

The left-hand side of (3.97) vanishes on solutions of the Euler–Lagrange equation, but is not identically zero, so ϕ is a first integral.

Example 3.15 In Example 3.14, we showed that $Q_1 = 1$ and $Q_2 = n$ are characteristics of variational point symmetries when

$$L = uu_1 - u^2.$$

Whenever L is a function of n, u and u_1 only, (3.98) amounts to

$$\phi = f - Q\,S_n^{-1}\,(L_{,1}).$$

Therefore, from (3.92) and (3.93), Q_1 and Q_2 yield the following first integrals of the Euler–Lagrange equation (3.91):

$$\phi^1 = u - u_{-1},$$
$$\phi^2 = (n - 1)u - nu_{-1}.$$

The non-point variational symmetries corresponding to $Q_4 = u_1 - u_{-1}$ yield

$$\phi^4 = u^2 - 2uu_{-1} + (u_{-1})^2 = \left(\phi^1\right)^2.$$

This illustrates an important point: there is no guarantee that different variational symmetries will produce functionally independent first integrals. ▲

Example 3.16 The McMillan map (3.82) in Example 3.13 has non-point
Lie symmetries whose characteristic is $Q = (u^2 + 1)(u_1 - u_{-1})$. Applying the
corresponding generator to the Lagrangian (3.83), one obtains

$$XL = (\mathrm{S}_n - \mathrm{I})\left\{u^2 + \left(u^2 - 2\alpha + 1\right)u_1 u_{-1}\right\},$$

so X generates variational symmetries. The resulting first integral is

$$\phi = u^2 + \left(u^2 - 2\alpha + 1\right)u_1 u_{-1} - Q\,\mathrm{S}_n^{-1}\frac{\partial L}{\partial u_1}$$

$$= u^2 - 2\alpha u_1 u_{-1} + \left(u^2 + 1\right)(u_{-1})^2.$$

The dynamical symmetry with $\widetilde{Q} = 2(u^2 + 1)u_1 - 2\alpha u$ does not generate variational symmetries. This seems odd, because $\widetilde{Q} - Q$ is zero if and only if u
is a solution of the McMillan map. So, in effect, the two characteristics produce the same symmetries of the Euler–Lagrange equation, but only one yields
variational symmetries. This demonstrates that two symmetries which have the
same effect on solutions of the Euler–Lagrange equation may, nevertheless, act
quite differently on the variational problem.   ▲

Notes and further reading

This chapter has introduced Lie groups of transformations in their simplest
context, in order to avoid too many technicalities. For a more thorough expo-
sition, with applications to differential equations, see Olver (1993). There are
many introductions to Lie algebras; for the 'pure' and 'applied' viewpoints,
consult Erdmann and Wildon (2006) and Fuchs and Schweigert (1997), respectively. Hydon (2000b) includes a simple explanation of why solvability is
needed if symmetries are to be inherited.

For scalar OΔEs, Lie point symmetries act on the dependent variable $u \in \mathbb{R}$.
Olver (1995) includes a proof that, subject to technical conditions, any finite-
dimensional Lie algebra of point transformations of a single complex variable,
u, can be written in suitable coordinates as one of $\mathrm{Span}(\partial_u)$, $\mathrm{Span}(\partial_u, u\partial_u)$
or $\mathrm{Span}(\partial_u, u\partial_u, u^2\partial_u)$. Olver also observes that the proof needs only minor
adjustments to apply to transformations of a real variable. Consequently, every
characteristic for an R-dimensional local Lie group of point symmetries can be
written in suitable local coordinates as $Q(n, u) = A(n) + B(n)u + C(n)u^2$.

Exercises

3.1   Find all solutions of (3.18) that are invariant under a one-parameter local
       Lie group of point symmetries. Which of these solutions belong to the
       general solution of (3.18)?

3.2 For the system of OΔEs

$$u_1 = (n+1)/v, \qquad v_1 = n/u, \qquad n \geq 1,$$

find all Lie symmetry generators $Q^1(n, u, v)\,\partial_u + Q^2(n, u, v)\,\partial_v$ such that each Q^i is linear in u and v. Use your results to solve the OΔEs.

3.3 Repeat Exercise 3.2 for the system

$$u_1 = v, \qquad v_1 = nu, \qquad n \geq 1.$$

3.4 Prove that every second-order linear homogeneous OΔE (3.35) has an eight-dimensional vector space of characteristics that are linear in u_1. Is this vector space a Lie algebra?

3.5 Show that if X_1 and X_2 commute, so do the group elements obtained by exponentiating $\varepsilon_1 X_1$ and $\varepsilon_2 X_2$.

3.6 Given a two-dimensional Lie algebra $\mathcal{L} = \mathrm{Span}(\tilde{X}_1, \tilde{X}_2)$, show that it is possible to choose a basis (X_1, X_2) such that either (3.56) or (3.57) holds.

3.7 Show that if X_i and X_j are Lie point symmetry generators for a given OΔE, so is $[X_i, X_j]$. Explain why the same is true for dynamical symmetry generators.

3.8 (a) Suppose that

$$Q_i = a_i(n, u)\, u_1 + b_i(n, u), \quad i = 1, 2,$$

are characteristics of dynamical symmetries for an OΔE of order $p \geq 2$, and that the function $a_1(n, u)$ is not identically zero. Under what conditions is $[Q_1, Q_2]$ linear in u_1?

(b) Show that for every linear OΔE, the set of all generators of the form

$$Q = a(n, u)\, u_1 + b(n, u)$$

is a Lie algebra.

3.9 (a) Find the general solution of the linear OΔE

$$u_2 - \pi_E(n)\, u_1 - u = 0, \qquad n \in \mathbb{Z},$$

by considering even and odd n separately. Hence calculate a pair of functionally independent first integrals.

(b) Use the above results to find all characteristics of dynamical symmetries that are linear in u_1.

(c) Now solve the LSC for Lie point symmetries and check that your results are consistent with those in (b). [This exercise is not difficult, but it contains a trap for the unwary!]

3.10 For the nonlinear OΔE

$$u_2 = 2u_1^3/u^2 - u_1,$$

show that every characteristic that is linear in u_1 is of the form

$$Q = c_1 u_1/u + c_2 u.$$

3.11 Determine a basis for the Lie algebra of (3.67), in each of the cases $k = 1, 2$ and 3.

3.12 Find all characteristics of symmetries of the linear OΔE

$$u_3 - 3u_2 + 3u_1 - u = 0$$

that are independent of u_2. Show that the corresponding generators form a ten-dimensional Lie algebra. Calculate the structure constants for this Lie algebra and show that it is not solvable.

3.13 Calculate the Lie algebra of Lie point symmetry generators for

$$u_3 = u_1 + \frac{1}{u_2 - u}.$$

Find a canonical basis and so determine the general solution of the OΔE.

3.14 Suppose that the space of all point symmetry characteristics for a given third-order scalar OΔE is spanned by $Q_1 = 1$, $Q_2 = u$ and $Q_3 = u^2$.

(a) Show that the OΔE is necessarily of the form

$$u_3 = \frac{uu_2 - g(n)\,u_1 u_2 - (1 - g(n))\,uu_1}{(1 - g(n))\,u_2 - u_1 + g(n)u}, \qquad g(n) \notin \{0, 1\}. \quad (3.99)$$

(b) Reduce (3.99) to a Riccati equation by writing it in terms of

$$z = \frac{u_2 - u_1}{u_1 - u},$$

which is invariant under the symmetries whose characteristics are Q_1 and Q_2. Hence show that the problem of solving (3.99) can be reduced to the problem of solving a linear second-order OΔE.

(c) Use the approach in (b) to find the general solution of

$$u_3 = \frac{4uu_2 - 3u_1u_2 - uu_1}{u_2 - 4u_1 + 3u}.$$

3.15 Calculate the Lie algebra of point symmetry generators for

$$u_3 = -\frac{u_1u_2 + uu_1 + 4}{uu_1u_2 + u_2 + u}.$$

Use them as appropriate to determine the general solution of the OΔE.

3.16 (a) Show that

$$u_2 = \frac{-uu_1^2 + 2u_1 - u}{u_1^2 - 2uu_1 + 1} \tag{3.100}$$

has a three-dimensional Lie algebra of point symmetry generators. Use these generators to linearize and solve (3.100).

(b) Find a way to linearize any non-partitioned p^{th}-order scalar OΔE that has $p + 1$ independent point symmetry generators (assuming that these have already been calculated).

3.17 Let Q_i, $i = 1, \ldots, R$, be linearly independent characteristics of point or dynamical symmetries of a given non-partitioned second-order OΔE. Define the functions

$$W_{ij} = Q_i \overline{S}_n Q_j - Q_j \overline{S}_n Q_i, \qquad 1 \le i < j \le R.$$

(a) Show that the ratio of any two linearly independent W_{ij} is a first integral.

(b) Show that if $W_{ij} = 0$ then Q_i/Q_j is a first integral.

(c) How can this approach be extended to higher-order OΔEs?

(d) To what extent do these results change if the OΔE is partitioned?

3.18 (a) Show that every *Möbius transformation*,

$$\hat{u} = \frac{\alpha(n)\, u + \beta(n)}{\gamma(n)\, u + \delta(n)}, \qquad \alpha(n)\delta(n) - \beta(n)\gamma(n) \ne 0,$$

maps the set of characteristics that are quadratic in u to itself.

(b) Show that if a Lie algebra $\mathcal{L}$ is not solvable but contains only characteristics that are quadratic in u, then $\mathcal{L}$ contains at least one characteristic Q that can be transformed to $\hat{Q} = 1$ by a real-valued Möbius transformation.

(c) Assuming that $\mathcal{L}$ contains the characteristic 1 and is not solvable, show that (after a linear point transformation if necessary) $\mathcal{L}$ contains $C_1(n)\,u^2 + A_1(n)$ where $C_1(n)$ is nonzero for all $n \in D$.

3.19 (a) Find the smallest Lie algebra $\mathcal{L}_{\min}$ that contains the characteristics

$$Q_1 = 1, \quad Q_2 = u, \quad Q_3 = u\sin(2n\pi/3).$$

(b) Are there Lie algebras of dimension six, seven and eight that contain $\mathcal{L}_{\min}$?

3.20 Find the most general third-order OΔE whose Lie point symmetry characteristics include $Q_1 = (-1)^n u$ and $Q_2 = u^2$.

3.21 (a) Find the smallest Lie algebra that contains the characteristics $Q_1 = 1$ and $Q_2 = (-1)^n u^2$.

(b) Use the above result to find a nonlinear OΔE whose Lie point symmetries include the one-parameter Lie groups that have the characteristics Q_1 and Q_2.

(c)* Generalize the result in (a) as follows. Suppose that the domain $\mathbb{Z}$ is regular for an OΔE whose characteristics include $Q_1 = 1$ and $Q_2 = A(n)\,u^2$. Determine the structure of the smallest Lie algebra that contains these characteristics, establishing all conditions that A must satisfy. Under what circumstances is this Lie algebra finite-dimensional? How are your results affected if the domain D is a finite interval, instead of $\mathbb{Z}$?

3.22 Show that if $L(n, u, u_1)$ satisfies (3.81) for all n in any finite interval, there exists a function $f(n, u)$ that satisfies (3.80).

3.23 (a) Show that, for any function $F(n, u, u_1, \dots)$ and any differential operator $X = Q\,\partial_u + S_n Q\,\partial_{u_1} + \cdots$, the following identity holds:

$$\mathbf{E}_u(XF) = \mathbf{E}_u(Q\mathbf{E}_u(F)). \tag{3.101}$$

(b) Hence establish that the condition (3.89) is sufficient to ensure that $\mathbf{E}_u(X^k L) \equiv 0$ for all $k \geq 2$. [Hint: use induction.]

(c) Show that if $Q = Q(n, u, \dots, u_l)$ then, for any $F(n, u, u_1, \dots)$,

$$[\mathbf{E}_u, X]F \equiv \mathbf{E}_u(X(F)) - X(\mathbf{E}_u(F)) = \sum_{k=0}^{l} S_n^{-k}(Q_{,k+1}\mathbf{E}_u(F)).$$

Use this result to prove that (3.89) guarantees that X generates symmetries of $\mathbf{E}_u(L) = 0$.

3.24 (a) Calculate the Euler–Lagrange equation for the variational problem
whose Lagrangian is

$$L = u/u_1 - \ln\left(u^2\right).$$

(b) Find all characteristics for point symmetries of the Euler–Lagrange
equation and show that the problem has variational symmetries.

(c) Use your results to construct a first integral; hence, solve the Euler–
Lagrange equation completely.

(d) Now try to solve the Euler–Lagrange equation by using canonical
coordinates corresponding to the Lie point symmetries. Is this approach easier than the variational symmetry method?

3.25 Calculate the Euler–Lagrange equation for

$$L = u_2/u + (-1)^n \ln|u| \ln|u_1|,$$

then reduce its order as far as can be achieved with variational Lie point
symmetries.

4

Lattice transformations

What is out of the common is usually a guide rather
than a hindrance. In solving a problem of this sort, the
grand thing is to be able to reason backwards.

(Sir Arthur Conan Doyle, A Study in Scarlet)

So far, we have focused mainly on symmetries that affect only the dependent
variables. These are very closely related to their counterparts for differential
equations. However, many of the distinctive features of difference equations
are seen only when one examines transformations of the independent variables.
This chapter explains the key ideas.

4.1 Transformations of the total space

We have seen various simple transformations of O∆Es and ODEs. More gen-
erally, a transformation of a space, $\mathcal{M}$, with local coordinates $\mathbf{z}$, is a bijection,

$$\Psi : \mathcal{M} \to \mathcal{M}, \qquad \Psi : \mathbf{z} \mapsto \hat{\mathbf{z}}, \tag{4.1}$$

with the extra property that $\mathcal{M}$ and its image, $\Psi(\mathcal{M})$, are identical. The term
'identical' means that not only do $\mathcal{M}$ and $\Psi(\mathcal{M})$ coincide as sets; they also
have exactly the same structure. The transformation is said to preserve an ob-
ject $\mathcal{P} \subset \mathcal{M}$ (which may have some extra structure) if $\mathcal{P}$ and $\Psi(\mathcal{P})$ are identical.

There are two complementary ways to view a transformation. The *active
view* is that $\mathbf{z}$ and $\hat{\mathbf{z}}$ represent two points in the given coordinate system. These
points are not necessarily distinct. For instance, the identity transformation
maps every point to itself; many other transformations have some fixed points.

By definition, $\mathcal{M}$ is preserved by Ψ, so the transformation can be treated as
a change of coordinates. Taking this *passive view*, each point is unchanged –
the transformation merely represents it in different coordinates, $\hat{\mathbf{z}}$. Both viewpoints are useful. To distinguish between coordinate systems, take the passive

view. To see how an object changes under a transformation, simply remove the carets; this yields the active view of the transformation.

For our purposes, the main objects of interest are difference equations and their solutions. We consider difference equations with N independent variables, $n^i \in \mathbb{Z}$, and M dependent variables, $u^\alpha \in \mathbb{R}$, writing $\mathbf{n} = (n^1, \dots, n^N)$ and $\mathbf{u} = (u^1, \dots, u^M)$. These variables can be regarded as coordinates on the *total space*, $\mathbb{Z}^N \times \mathbb{R}^M$. However, as far as a given difference equation is concerned, they may only be local coordinates; several different coordinate patches might be needed to describe the whole difference equation[1].

To avoid a proliferation of indices, we continue to use u as the dependent variable in any scalar equation and n as the independent variable in any OΔE or system of OΔEs. For partial difference equations (PΔEs) with $N = 2$, we use (m, n) rather than (n^1, n^2). These variables are free to take any values in a given domain, $D \subset \mathbb{Z}^2$; it is convenient to suppress them, using u and $u_{i,j}$ to denote $u(m, n)$ and $u(m+i, n+j)$, respectively. The alternative notation $u_{0,0}$ is occasionally used in place of u.

The simplest kind of transformation of a given difference equation arises from a transformation of the total space,

$$\Psi : \mathbb{Z}^N \times \mathbb{R}^M \to \mathbb{Z}^N \times \mathbb{R}^M, \qquad \Psi : (\mathbf{n}, \mathbf{u}) \mapsto (\hat{\mathbf{n}}, \hat{\mathbf{u}}). \qquad (4.2)$$

The following example illustrates the active and passive views in this context.

Example 4.1 Consider a transformation (4.2) with $N = 2, M = 1$ and

$$(\hat{m}, \hat{n}, \hat{u}) = (m + 2n, \; n, \; 4u + m - n).$$

From the active viewpoint, working in the (m, n, u) coordinate system, the transformation maps $(1, 2, 0.5)$ to $(5, 2, 1)$. The passive view is that the point whose (m, n, u) coordinates are $(1, 2, 0.5)$ is represented in the $(\hat{m}, \hat{n}, \hat{u})$ coordinate system by $(5, 2, 1)$. ▲

We now examine the structure of the total space, in order to identify which maps Ψ preserve this space and so produce valid coordinate changes. The most obvious structural feature is seen in Figure 4.1: the total space is a *lattice* $(\mathbb{Z}^N)$ with a *fibre* $(\mathbb{R}^M)$ over each lattice point, $\mathbf{n}$. Fibres are characterized by

$$\pi : \mathbb{Z}^N \times \mathbb{R}^M \to \mathbb{Z}^N, \qquad \pi : (\mathbf{n}, \mathbf{u}) \mapsto \mathbf{n},$$

which maps every point on a given fibre to a single lattice point. A change of

---

[1] This applies, for instance, to a finite difference approximation of fluid flow in a domain that is not simply connected.

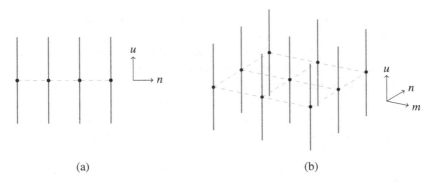

Figure 4.1 Solutions of scalar difference equations lie on $\mathbb{Z}^N \times \mathbb{R}$, where N is the number of independent variables: (a) $N = 1$; (b) $N = 2$.

coordinates, from $(\mathbf{n}, \mathbf{u})$ to $(\hat{\mathbf{n}}, \hat{\mathbf{u}})$, is compatible with the projection π (which means that it maps each fibre to a fibre) if and only if Ψ is of the form

$$\Psi : (\mathbf{n}, \mathbf{u}) \mapsto (\hat{\mathbf{n}}, \hat{\mathbf{u}}) = (\nu(\mathbf{n}), \psi(\mathbf{n}, \mathbf{u})). \tag{4.3}$$

Although this condition is necessary, it is not sufficient to preserve the structure of total space. To see what further conditions apply, we will look at coordinate changes on the lattice and on each fibre separately. This approach is justified by the splitting $\Psi = \Psi_2 \circ \Psi_1$, where

$$\Psi_1 : (\mathbf{n}, \mathbf{u}) \mapsto (\hat{\mathbf{n}}, \mathbf{u}) = (\nu(\mathbf{n}), \mathbf{u}),$$
$$\Psi_2 : (\hat{\mathbf{n}}, \mathbf{u}) \mapsto (\hat{\mathbf{n}}, \hat{\mathbf{u}}) = \left(\hat{\mathbf{n}}, \psi(\nu^{-1}(\hat{\mathbf{n}}), \mathbf{u})\right).$$

4.1.1 Lattice maps

The change of independent variables is determined by ν, which we call the *lattice map*. We describe this bijection as *valid* if it preserves the structure of the lattice. A valid lattice map cannot shuffle lattice points arbitrarily, because the image of the lattice must be the Cartesian product of N copies of $\mathbb{Z}$. Each discrete line ($\mathbb{Z}$) is ordered, enabling the forward and backward shift operators to be defined. To recognize which lattice maps preserve these features (and are therefore valid coordinate changes), we must consider the relationships between lattice points.

It is helpful to associate lattice points with their position vectors from a fixed origin, $\mathbf{0}$. Let $\mathbf{e}_i$ denote the vector with $n^i = 1$ and $n^j = 0$ for all $j \neq i$; these vectors form a basis, so each point in the lattice can be written as $\mathbf{n} = n^i \mathbf{e}_i$.

The point $\mathbf{n} + \mathbf{m}$ is *visible* from $\mathbf{n}$ (and vice versa) if $\mathbf{n} + \lambda\mathbf{m} \notin \mathbb{Z}^N$ for all (real) $\lambda \in (0, 1)$, that is, if there are no lattice points between $\mathbf{n}$ and $\mathbf{n} + \mathbf{m}$.

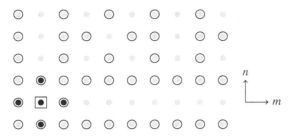

Figure 4.2 The point enclosed by a square is adjacent to the other black points; it is visible from each ringed point, but not from any unringed point.

These two points are mutually visible if and only if the components of **m** are coprime (that is, their greatest common divisor is 1). It is important to realize that visibility has nothing to do with closeness in the usual sense. For instance, the point $(99, 98)$ is visible from $(0, 0)$ and $(2, 0)$, but $(0, 0)$ and $(2, 0)$ are not mutually visible.

We reserve the term *adjacent* for two (mutually visible) points that differ by $\pm \mathbf{e}_i$ for some i. Thus, each point in $\mathbb{Z}^N$ is adjacent to exactly $2N$ points; see Figure 4.2. Adjacency is defined relative to a given coordinate system, so a coordinate change may affect which pairs of points are adjacent. This leads to a useful technique for solving or simplifying some PΔEs (see §4.3). In some circumstances, however, it is necessary to use only those transformations that preserve adjacency.

If **n** is visible from $\mathbf{n} + \mathbf{m}$, it is also visible from $\mathbf{n} - \mathbf{m}$. These points lie on an ordered (infinite) line, in which each point is visible from its two immediate neighbours:

$$\ell_{\mathbf{n}}(\mathbf{m}) = (\ldots, \mathbf{n} - 2\mathbf{m}, \ \mathbf{n} - \mathbf{m}, \ \mathbf{n}, \ \mathbf{n} + \mathbf{m}, \ \mathbf{n} + 2\mathbf{m}, \ldots). \tag{4.4}$$

The direction of this line depends only on **m**; all lines with the same **m** are parallel. Any valid coordinate change will map each line (4.4) to a line, preserving parallelism (and visibility); therefore v must be an invertible affine transformation[2] of the lattice, as follows.

Let $\mathbf{n}_0 = v(\mathbf{0})$ and (with a slight abuse of notation) let $\hat{\mathbf{e}}_i = v(\mathbf{e}_i) - \mathbf{n}_0$. Each $\mathbf{e}_i$ is visible from **0** and every other $\mathbf{e}_j$, so each pair of points in $\{\mathbf{0}, \hat{\mathbf{e}}_1, \ldots, \hat{\mathbf{e}}_N\}$ is mutually visible. The requirement that v is affine determines $v(\mathbf{n})$ everywhere:

$$v(\mathbf{n}) = n^i \hat{\mathbf{e}}_i + \mathbf{n}_0. \tag{4.5}$$

---

[2] An affine transformation can be regarded as a linear transformation followed by a translation.

In particular, $\ell_0(\mathbf{e}_i)$ is mapped to $\ell_{\mathbf{n}_0}(\hat{\mathbf{e}}_i)$ for each i, and the Cartesian product $\ell_{\mathbf{n}_0}(\hat{\mathbf{e}}_1) \times \cdots \times \ell_{\mathbf{n}_0}(\hat{\mathbf{e}}_N)$ is the entire lattice.

Let $\hat{\mathbf{e}}_i = a_i^j \mathbf{e}_j$; by taking the active view and leaving the basis vectors $\mathbf{e}_j$ fixed, one can regard v as a map on the components of $\mathbf{n}$:

$$v : n^j \mapsto a_i^j n^i + n_0^j. \tag{4.6}$$

As this map is bijective, the matrix $A = (a_i^j)$ is non-singular and A^{-1} also has integer components. Therefore A is a *unimodular matrix*, that is, it belongs to the general linear group $\mathrm{GL}_N(\mathbb{Z})$, which is the group of $N \times N$ matrices with integer coefficients and determinant ± 1. There are no further constraints on v; it is valid if and only if A is unimodular. The special linear group $\mathrm{SL}_N(\mathbb{Z})$ is the subgroup of $\mathrm{GL}_N(\mathbb{Z})$ that consists of matrices whose determinant is 1. So if A is unimodular then either $A \in \mathrm{SL}_N(\mathbb{Z})$ or A is the composition of the reflection matrix $R = \mathrm{diag}(-1, 1, \ldots, 1)$ with a matrix from $\mathrm{SL}_N(\mathbb{Z})$.

A valid lattice map is rigid if it preserves adjacency, which (for $N \geq 2$) is a stronger requirement than the preservation of visibility. Clearly, every translation is rigid. Now consider which valid lattice maps that fix the origin are rigid. Adjacency is preserved if each $\mathbf{e}_i$ maps to $\pm \mathbf{e}_{j_i}$ (for some j_i), so A has just one nonzero entry in each row and column; each such entry is either 1 or -1. Consequently, every rigid lattice map that fixes the origin preserves the scalar product, $\mathbf{a} \cdot \mathbf{b}$, for all $\mathbf{a}, \mathbf{b} \in \mathbb{Z}^N$.

Example 4.2 The group $\mathrm{SL}_2(\mathbb{Z})$ is generated by

$$S = \begin{bmatrix} 0 & -1 \\ 1 & 0 \end{bmatrix}, \quad T = \begin{bmatrix} 1 & 1 \\ 0 & 1 \end{bmatrix}, \quad S^4 = S^{-1}TSTST = I. \tag{4.7}$$

Here S is a rotation by $\pi/2$ (which is rigid), T is a shear (which is not rigid), and I is the identity matrix. The group of rigid transformations of $\mathbb{Z}^2$ is generated by S, the reflection $R = \mathrm{diag}(-1, 1)$, and the translations v_1 and v_2.

Similarly, for each N, the group of rigid lattice maps is generated by a rotation by $\pi/2$ in each plane that has a basis $(\mathbf{e}_i, \mathbf{e}_j)$, the reflection whose matrix is R, and the translations v_i. ▲

For many difference equations, the independent variables do not take all possible values in $\mathbb{Z}^N$. When $N = 1$, the domain is a set of consecutive points in $\mathbb{Z}$; such a set is called a *discrete interval*, whether or not it is finite. The simplest N-dimensional analogue of a discrete interval is the Cartesian product of N discrete intervals, which is called a *box*. However, under a non-rigid change of coordinates, a box will appear to be deformed. To overcome this difficulty, we define a *product lattice* to be any set of points in $\mathbb{Z}^N$ that can be mapped to a

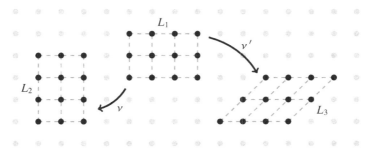

Figure 4.3 Valid lattice maps acting on a box, L_1: v is rigid; v' is not rigid.

box by a valid lattice map. In Figure 4.3, for instance, the sets L_1, L_2 and L_3 are product lattices, but only L_1 and L_2 are boxes.

4.1.2 Fibre-preserving transformations

Given a differential equation, any transformation of the independent and dependent variables only is a point transformation. This has an analogue for difference equations, using local coordinates $(\mathbf{n}, \mathbf{u})$, where $\mathbf{n}$ takes all values within a product lattice and $\mathbf{u}$ coordinatizes an open subset of the fibre over $\mathbf{n}$. Every transformation of $\mathbf{n}$ and $\mathbf{u}$ only is (locally) a fibre-preserving transformation, that is, a mapping of the form

$$\Psi : (\mathbf{n}, \mathbf{u}) \mapsto (\hat{\mathbf{n}}, \hat{\mathbf{u}}) = (v(\mathbf{n}), \psi_{\mathbf{n}}(\mathbf{u})), \tag{4.8}$$

where v is a valid lattice map and $\psi_{\mathbf{n}} = \psi(\mathbf{n}, \cdot)$ is a point transformation for each $\mathbf{n}$.

Suppose that an object, $\mathcal{P}$, is described in terms of $\mathbf{n}$ and $\mathbf{u}$ by $P(\mathbf{n}, \mathbf{u})$. Taking the passive view, it is also described by

$$\tilde{P}(\hat{\mathbf{n}}, \hat{\mathbf{u}}) = P\left(v^{-1}(\hat{\mathbf{n}}), (\psi_{v^{-1}(\hat{\mathbf{n}})})^{-1}(\hat{\mathbf{u}})\right). \tag{4.9}$$

Thus, in the active view, Ψ transforms $\mathcal{P}$ to $\tilde{\mathcal{P}}$, which is represented in $(\mathbf{n}, \mathbf{u})$ coordinates by $\tilde{P}(\mathbf{n}, \mathbf{u})$.

4.1.3 Lattice transformations of a PΔE

Difference equations are described in terms of $\mathbf{n}, \mathbf{u}$ and shifts of $\mathbf{u}$. Therefore, to see how a given difference equation is transformed, one must prolong the transformation Ψ. We have already seen how this is done for Lie point symmetries of OΔEs; for the remainder of this chapter, we will look at the effect of transforming $\mathbf{n}$.

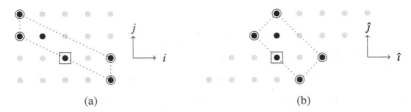

Figure 4.4 (a) The stencil of (4.12). (b) The stencil of (4.13). In each case, the stencil is the set of black points; every vertex is ringed, dotted lines indicate the boundary and the point $(0, 0)$ is enclosed by a square.

In the particular case $\psi_{\mathbf{n}}(\mathbf{u}) = \mathbf{u}$, we call Ψ a *lattice transformation*. Replacing $\mathbf{n}$ by $\mathbf{n} + \mathbf{i}$ yields the prolongation of a lattice transformation:

$$(\hat{\mathbf{n}}, \hat{\mathbf{u}}, \hat{\mathbf{u}}_{\hat{\mathbf{i}}}) = (v(\mathbf{n}), \mathbf{u}, \mathbf{u}_{\mathbf{i}}), \qquad \text{where } \hat{\mathbf{i}} = v(\mathbf{n} + \mathbf{i}) - v(\mathbf{n}). \qquad (4.10)$$

As v is affine, $\hat{\mathbf{i}}$ depends on $\mathbf{i}$, but not explicitly on $\mathbf{n}$.

In Chapter 1, we saw that translations and reflections may transform a given OΔE to a known OΔE of the same order; they can also be used to simplify a given PΔE. Some non-rigid lattice transformations can do more: they change the (apparent) orders of a PΔE, as we now describe.

4.2 What are the orders of a given PΔE?

For brevity, we consider only scalar PΔEs whose domain is $D \subset \mathbb{Z}^2$ for the rest of this chapter[3]. Such a PΔE is an expression of the form

$$\mathcal{A}(m, n, \{u_{i,j} : (i, j) \in \mathcal{S}\}) = 0, \qquad (m, n) \in D. \qquad (4.11)$$

Here $\mathcal{S}$ is a non-empty finite set, called the *stencil* (or mask) of the PΔE; usually, we write $u_{0,0}$ as u. Given a PΔE in the (m, n) coordinate system, one finds the stencil by identifying each pair (i, j) such that $u_{i,j}$ occurs in the PΔE. For instance, Figure 4.4(a) depicts the stencil for

$$u_{2,0} - (u_{-2,2} - 2u_{-1,1} + u)(u_{2,-1} - u_{-2,1}) = 0, \qquad m, n \geq 1. \qquad (4.12)$$

The stencil of a PΔE may be translated in the (i, j)-plane by rewriting (4.11) as

$$\mathcal{A}(m+k, n+l, \{u_{i+k,j+l} : (i, j) \in \mathcal{S}\}) = 0, \qquad (m+k, n+l) \in D.$$

This translates the stencil by (k, l); in compensation, the domain is translated by $(-k, -l)$.

---

[3] The main ideas apply equally to PΔEs with more than two independent variables and to systems of PΔEs. However, the notation becomes messy!

A *border* is a line in the (i, j)-plane, $\ell_i(\mathbf{k})$, that includes at least two points of $\mathcal{S}$ and is not straddled by $\mathcal{S}$. Consequently, the whole stencil lies on the same side of $\ell_i(\mathbf{k})$. The *boundary* of the stencil is the set of points in $\mathcal{S}$ that belong to at least one border; all other points in $\mathcal{S}$ are interior points. A *vertex* is a point that belongs to two distinct borders, so each border contains exactly two vertices. The set of vertices[4] is denoted by $V_\mathcal{S}$.

The *order* of the P∆E in m (resp. n) is the difference between the greatest and least values of i (resp. j) in the stencil. As the vertices lie at the extremes of the stencil, the order in each direction can be calculated from $V_\mathcal{S}$ alone. However, although $V_\mathcal{S}$ is a coordinate-independent object, the orders of the P∆E are defined with respect to particular coordinates. Consequently, the orders may be changed by a lattice transformation, (4.10), with $\hat{\mathbf{i}} = (\hat{\imath}, \hat{\jmath})$.

Example 4.3   The P∆E (4.12) is fourth-order in m and third-order in n. The unit shear in the m-direction, $(\hat{m}, \hat{n}, \hat{u}) = (m+n, n, u)$, transforms (4.12) into

$$\hat{u}_{2,0} - (\hat{u}_{0,2} - 2\hat{u}_{0,1} + \hat{u})(\hat{u}_{1,-1} - \hat{u}_{-1,1}) = 0, \qquad 1 \le \hat{n} < \hat{m}, \qquad (4.13)$$

which is third-order in each of $\hat{m}$ and $\hat{n}$. Figure 4.4(b) shows this transformation from the passive viewpoint.   ▲

Lemma 4.4   *Every lattice transformation maps vertices to vertices, boundary points to boundary points, and interior points to interior points.*

Proof   From the passive viewpoint, a lattice transformation changes the coordinates on the (i, j)-plane. The definitions of vertices, boundary points and interior points are coordinate-independent.   □

Lemma 4.5   *For any given vertex, V, there exists a lattice transformation such that the $\hat{\imath}$ coordinate value of V exceeds that of every other vertex, making V the rightmost stencil point in the $(\hat{\imath}, \hat{\jmath})$-plane.*

Proof   Assume that V is located at the origin in the (i, j)-plane. (If necessary, translate the stencil to achieve this.) Using the notation (4.4), the two borders that intersect at V are $\ell_0(\mathbf{k}_1)$ and $\ell_0(\mathbf{k}_2)$. Here each $\mathbf{k}_\alpha = (k_\alpha, l_\alpha)^\mathrm{T}$ is visible from $\mathbf{0}$ and points towards the other vertex that shares the border with V (see Figure 4.5); the indices are chosen so that $k_1 l_2 - k_2 l_1 > 0$.

If $\mathbf{k}_1 \cdot \mathbf{k}_2 > 0$, the boundary of the stencil has an acute interior angle at V. In this case, the required transformation is rotation by the integer multiple of $\pi/2$ that yields $\hat{k}_1 < 0$, $\hat{l}_1 \ge 0$.

---

[4]   This set generates the convex hull of the stencil, which is the smallest convex lattice polygon that contains $\mathcal{S}$.

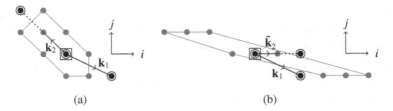

Figure 4.5 (a) In the original configuration, $\mathbf{k}_1 = (2, -1)^{\mathrm{T}}$ and $\mathbf{k}_2 = (-1, 1)^{\mathrm{T}}$. (b) The shear transformation (4.14) (with $q = 1$), using the active viewpoint. The symbols are as in Figure 4.4; the grey lines are added for clarity.

If $\mathbf{k}_1 \cdot \mathbf{k}_2 \leq 0$, a shear must first be used to create an acute interior angle, as shown in Figure 4.5. The matrix

$$A = \begin{bmatrix} 1 - k_1 l_1 & (k_1)^2 \\ -(l_1)^2 & 1 + k_1 l_1 \end{bmatrix} \tag{4.14}$$

gives a unit shear in the $\mathbf{k}_1$-direction. Applying this q times yields

$$\mathbf{k}_1 \mapsto \mathbf{k}_1, \qquad \mathbf{k}_2 \mapsto \tilde{\mathbf{k}}_2 = \mathbf{k}_2 + q(k_1 l_2 - k_2 l_1)\mathbf{k}_1.$$

The transformed angle is acute when $\mathbf{k}_1 \cdot \tilde{\mathbf{k}}_2 > 0$, that is, when

$$k_1 k_2 + l_1 l_2 + q(k_1 l_2 - k_2 l_1)\left((k_1)^2 + (l_1)^2\right) > 0.$$

This inequality holds for all sufficiently large q, so all that remains is to rotate the coordinates, as the first case. From Lemma 4.4, the vertices are the only points in $\mathcal{S}$ that can be made rightmost. □

Similarly, for a given border, there exists a lattice transformation that makes the border rightmost, which means that the transformed border is a line $\hat{\imath} = i_0$ and that every point in the transformed stencil has $\hat{\imath} \leq i_0$. (The proof of this is left as an exercise.)

Given a PΔE, $\mathcal{A} = 0$, the *stencil at* (m, n), denoted $\mathcal{S}(m, n)$, is obtained by evaluating $\mathcal{A}$ at (m, n) and finding the stencil of the resulting expression. Clearly, the PΔE can hold only if u is defined at every point of $\mathcal{S}(m, n)$, for each $(m, n) \in D$. Thus, solutions are defined on the *solution domain*,

$$D^+ = \bigcup_{(m,n)\in D} \mathcal{S}(m, n). \tag{4.15}$$

Each solution may be regarded as a graph,

$$\mathcal{G}_f = \{(m, n, f(m, n)) : (m, n) \in D^+\}, \tag{4.16}$$

with the property that $u = f(m, n)$ (prolonged as necessary) satisfies the PΔE for each $(m, n) \in D$. Note that the solution domain D^+ and the graph $\mathcal{G}_f$ are each unaffected by any translation of the stencil.

By definition, $\mathcal{S}(m, n) \subset \mathcal{S}$; typically, $\mathcal{S}(m, n)$ coincides with $\mathcal{S}$. Any point (m, n) at which at least one vertex of $\mathcal{S}$ does not belong to $\mathcal{S}(m, n)$ is called a *singular point*.

Example 4.6   The stencil of the linear PΔE

$$u_{2,2} + (m + n) \, u_{1,1} - (m - n + 1) \, u_{0,1} + \left(m^2 + n^2 \right) u = 0, \qquad (m, n) \in \mathbb{Z}^2, \quad (4.17)$$

is $\mathcal{S} = \{(0,0), (0,1), (1,1), (2,2)\}$; of these points, only $(1, 1)$ is not a vertex. The stencils $\mathcal{S}(m, n)$ that do not coincide with $\mathcal{S}$ are as follows:

For $m \neq 0$, $\mathcal{S}(m, -m) = \{(0,0), (0,1), (2,2)\}$; $\mathcal{S}(0,0) = \{(0,1), (2,2)\}$;
$$\mathcal{S}(m, m + 1) = \{(0,0), (1,1), (2,2)\}.$$

So $(0, 0)$ and every $(m, m+1)$ are the singular points for the PΔE.   ▲

Example 4.6 may seem somewhat counter-intuitive. A singular point of an OΔE is a point where the order of the OΔE is lowered. Clearly, the order in n is reduced at $(0, 0)$, but the PΔE (4.17) remains second-order in both m and n on the line $\{(m, m+1) : m \in \mathbb{Z}\}$. The resolution of this apparent difficulty is as follows. At a singular point, the PΔE loses order in some coordinate systems, including all systems in which a missing vertex is the rightmost vertex in $\mathcal{S}$. Lemma 4.5 guarantees that such a coordinate system exists for every vertex. In other words, the PΔE is degenerate at every singular point, though this degeneracy may be hidden by the choice of coordinate system.

For instance, in the coordinates $(\hat{m}, \hat{n}) = (n - m, n)$, the PΔE (4.17) is

$$\hat{u}_{0,2} + (2\hat{n} - \hat{m}) \, \hat{u}_{0,1} + (\hat{m} - 1) \, \hat{u}_{1,1} + \left(2\hat{n}^2 - 2\hat{m}\hat{n} + \hat{m}^2 \right) \hat{u} = 0. \qquad (4.18)$$

The line of singular points $(m, m+1)$ amounts to $\hat{m} = 1$, on which (4.18) ceases to be first-order in $\hat{m}$. Therefore $\hat{u}(1, \hat{n}) = v(\hat{n})$, where v satisfies the OΔE

$$v_2 + (2\hat{n} - 1) \, v_1 + \left(2\hat{n}^2 - 2\hat{n} + 1 \right) v = 0.$$

A PΔE does not lose order (in any coordinate system) at any point where $\mathcal{S}(m, n) \neq \mathcal{S}$ but $V_{\mathcal{S}(m,n)} = V_{\mathcal{S}}$, because $V_{\mathcal{S}(m,n)}$ alone is sufficient to determine the orders. This is why the definition of a singular point refers only to vertices.

Mirroring the definition for OΔEs, we define a *regular domain* for a given PΔE to be a product lattice that has no singular points. Some PΔEs are defined on domains that are free of singular points, but are not product lattices. It is possible to deal with such domains, but that is beyond the scope of this book.

4.3 Minimization of order

Given a PΔE on a regular domain, one can always choose a coordinate system $(\hat{m}, \hat{n})$ in which the PΔE has the lowest possible order in at least one variable. The way to do this is as follows. For definiteness, we minimize the order of the PΔE in $\hat{m}$ first. Any remaining freedom is then used to minimize the order in $\hat{n}$ as much as is possible without increasing the order in $\hat{m}$.

Both orders are unaffected by reflection in any line of constant m or n and by translations. So we consider only transformations of the form

$$\begin{pmatrix} \hat{m} \\ \hat{n} \end{pmatrix} = A \begin{pmatrix} m \\ n \end{pmatrix}, \qquad A = \begin{pmatrix} a & b \\ c & d \end{pmatrix}, \tag{4.19}$$

where $A \in SL_2(\mathbb{Z})$, with $a \geq 0$. (If $a = 0$, it is sufficient to consider $b > 0$; as $A \in SL_2(\mathbb{Z})$, this implies that $b = 1$.) The process of finding the transformations that minimize the orders of a given PΔE has two stages[5].

I. Find all pairs $(a, b) \in \mathbb{Z}^2$ that minimize the difference between the greatest and least of the numbers $ai + bj$, $(i, j) \in V_S$, subject to either $a > 0$ or $(a, b) = (0, 1)$. Such pairs will be coprime.

II. For each such (a, b), find all pairs $(c, d) \in \mathbb{Z}^2$ that minimize the maximum difference between the numbers $ci + dj$, subject to the constraint $ad - bc = 1$.

Example 4.7   Reduce the orders of the PΔE (4.12) as far as possible.

Solution: The vertices of the stencil for (4.12) are $(2, 0)$, $(2, -1)$, $(-2, 1)$ and $(-2, 2)$. First, minimize the maximum difference between $2a$, $2a - b$, $-2a + b$ and $-2a + 2b$, subject to either $a > 0$ or $(a, b) = (0, 1)$. This yields a single solution, $(a, b) = (1, 2)$. The maximum difference is 2, so (from the active viewpoint) the transformed PΔE will be second-order in m.

The condition $ad - bc = 1$ amounts to $d = 1 + 2c$, so all that remains is to minimize the maximum difference between $2c$, -1, 1 and $2c + 2$. There are two solutions, $c = 0$ and $c = -1$, each of which gives a maximum difference of 3. To summarize, either of the special linear matrices

$$A_1 = \begin{pmatrix} 1 & 2 \\ 0 & 1 \end{pmatrix} = T^2, \qquad A_2 = \begin{pmatrix} 1 & 2 \\ -1 & -1 \end{pmatrix} = T^{-1}S^{-1}T$$

can be used to transform (4.12) into a PΔE that is second-order in $\hat{m}$ and third-order in $\hat{n}$. Specifically, the transformation corresponding to A_1 yields the PΔE

$$\hat{u}_{2,0} - (\hat{u}_{2,2} - 2\hat{u}_{1,1} + \hat{u})(\hat{u}_{0,-1} - \hat{u}_{0,1}) = 0, \qquad 1 \leq \hat{n}, \quad 2\hat{n} + 1 \leq \hat{m};$$

---

[5] Each stage is an integer linear programming problem; such problems are NP-complete. However, most interesting PΔEs with two independent variables have a stencil of only three or four vertices, in which case the minimization is easily done by hand.

the transformation corresponding to A_2 gives

$$\hat{u}_{2,-2} - (\hat{u}_{2,0} - 2\hat{u}_{1,0} + \hat{u})(\hat{u}_{0,-1} - \hat{u}_{0,1}) = 0, \quad 1 - \hat{n} \leq \hat{m} \leq -(2\hat{n} + 1). \quad \blacktriangle$$

Some PΔEs can be solved completely, because they have a general solution that can be expressed in terms of arbitrary functions, which may be determined if initial conditions are given. These PΔEs are solved by slightly modified OΔE methods. The simplest case occurs when a given PΔE can be transformed to a parametrized OΔE by a non-rigid lattice transformation; other cases are treated in Chapter 5. Once the OΔE is solved, the general solution of the PΔE is obtained by replacing arbitrary constants by arbitrary functions of $\hat{m}$.

Example 4.8 Solve the linear PΔE

$$u_{2,4} - 2u_{1,2} + u = 0. \tag{4.20}$$

Solution: The minimization problem has infinitely many solutions,

$$A = \begin{pmatrix} 2 & -1 \\ 1 + 2d & -d \end{pmatrix}, \quad d \in \mathbb{Z}.$$

The simplest of these gives $(\hat{m}, \hat{n}) = (2m - n, m)$. With this change of variables, the PΔE (4.20) amounts to a second-order parametrized OΔE,

$$\hat{u}_{0,2} - 2\hat{u}_{0,1} + \hat{u} = 0,$$

whose general solution is $\hat{u} = \hat{n} g(\hat{m}) + h(\hat{m})$, for arbitrary functions g and h. Therefore, the general solution of (4.20) is

$$u = m g(2m - n) + h(2m - n). \quad \blacktriangle$$

Example 4.9 Solve the initial-value problem

$$u_{0,1} = \frac{u_{1,0}}{1 + n u_{1,0}}, \quad n \geq 1, \quad u(m, 1) = \frac{2}{m^2 + 1}. \tag{4.21}$$

Solution: Applying the minimization process gives

$$A = \begin{pmatrix} 1 & 1 \\ c & 1 + c \end{pmatrix}, \quad c \in \mathbb{Z}.$$

For simplicity, let $(\hat{m}, \hat{n}) = (m + n, n)$; then the initial-value problem (4.21) amounts to

$$\hat{u}_{1,1} = \frac{\hat{u}_{1,0}}{1 + \hat{n}\hat{u}_{1,0}}, \quad \hat{n} \geq 1, \quad \hat{u}(\hat{m}, 1) = \frac{2}{(\hat{m} - 1)^2 + 1}. \tag{4.22}$$

From Example 2.8, the general solution of the parametrized OΔE in (4.22) is

$$\hat{u} = \frac{2}{\hat{n}(\hat{n} - 1) + g(\hat{m})},$$

where g is an arbitrary function. Applying the initial condition gives

$$g(\hat{m}) = (\hat{m} - 1)^2 + 1.$$

Therefore the solution of the initial-value problem (4.21) is

$$u = \frac{2}{n(n-1) + (m+n-1)^2 + 1}. \qquad \blacktriangle$$

In each of the last two examples, there is an infinite family of minimizing transformations; any integer multiple of the parameter $\hat{m}$ may be added to $\hat{n}$. This freedom occurs whenever a given PΔE can be reduced to a parametrized OΔE. The stencil lies on a single line, so the reduction is unaffected by an arbitrary shear in the direction of that line.

So far, we have sought to minimize the order in $\hat{m}$ as far as possible, without considering whether or not the transformed PΔE has a rightmost vertex. Sometimes, it is useful for a particular vertex, $\mathcal{V}$, to be rightmost (see §4.4). Let (i_0, j_0) be the stencil coordinates of $\mathcal{V}$ with respect to (m, n). To make $\mathcal{V}$ rightmost while keeping the order of the PΔE in $\hat{m}$ as small as is then possible, replace Stage I in the usual minimization process by the following, leaving Stage II unchanged.

I$'$. Find every pair $(a, b) \in \mathbb{Z}^2$ that minimizes the maximum of the numbers $a(i_0-i) + b(j_0-j)$, $(i, j) \in V_S \setminus \{(i_0, j_0)\}$, subject to each of these numbers being positive. Such pairs will be coprime.

Example 4.10 The stencil for the PΔE

$$u_{1,-1} + u_{0,1} + u_{-1,0} - (u_{2,1} - u_{0,-1})^2 = 0 \qquad (4.23)$$

is

$$S = \{(1, -1), (0, 1), (-1, 0), (2, 1), (0, -1)\}; \qquad (4.24)$$

here, every stencil point is a vertex. Find an $\mathrm{SL}_2(\mathbb{Z})$ transformation that minimizes the orders, subject to the vertex at $(1, -1)$ becoming rightmost.

Solution: For Stage I$'$, we need to minimize the maximum of

$$a - 2b, \quad 2a - b, \quad -a - 2b, \quad a,$$

given that each of these numbers is positive. The positivity constraints require that $0 < a < -2b$; hence, the minimizing solution is $(a, b) = (1, -1)$. Stage II works as usual and gives a single solution, $(c, d) = (0, 1)$. So the transformed PΔE is

$$\hat{u}_{2,-1} + \hat{u}_{-1,1} + \hat{u}_{-1,0} - (\hat{u}_{1,1} - \hat{u}_{1,-1})^2 = 0,$$

which is third-order in $\hat{m}$ and second-order in $\hat{n}$. $\qquad \blacktriangle$

In the above example, the minimizing transformation happened to yield a PΔE of the same orders as the original PΔE. This is unusual. For instance, the minimizing transformation of (4.23) that makes $(0, 1)$ the rightmost vertex yields a PΔE that is fifth-order in $\hat{m}$ and second-order in $\hat{n}$ (see Exercise 4.10).

4.4 Initial-value problems for scalar PΔEs

In the simplest type of initial-value problem (IVP), one is given a scalar PΔE (4.11) with a domain of the form $D = \{(m, n) \in \mathbb{Z}^2 : m \geq m_0\}$. If the PΔE is of order p in m, the problem is to find u throughout D, for given initial conditions $\{u(m, n) : m = m_0, \ldots, m_0 + p - 1, \ n \in \mathbb{Z}\}$.

Theorem 4.11 *Suppose that the scalar PΔE in the above IVP is*

$$u_{p,J} = \omega\left(m, n, \mathbf{u}_{\{0,\cdot\}}, \mathbf{u}_{\{1,\cdot\}}, \ldots, \mathbf{u}_{\{p-1,\cdot\}}\right), \tag{4.25}$$

where $\mathbf{u}_{\{i,\cdot\}}$ is shorthand for a given finite subset of $\{u_{i,j} : j \in \mathbb{Z}\}$. If, for each $(m, n) \in D$, the right-hand side of (4.25) is finite for all $u_{i,j}$, the IVP has a unique solution for arbitrary initial conditions.

Proof The solution can be constructed recursively, on one line of constant m at a time, with each $u(m, n)$ being determined uniquely by a set of values that are already known. □

We call (4.25) *Kovalevskaya form*, by analogy with the corresponding form for PDEs. The term $u_{p,J}$ is isolated, that is, $u_{p,J}$ appears nowhere else in the PΔE. It is called the *leading term*; correspondingly, (p, J) is the *leading vertex*. (One can set $J = 0$ by translating in n.) Usually, it is obvious whether or not a given PΔE can be written in Kovalevskaya form, but sometimes a little work is needed to decide this.

Example 4.12 Consider the PΔE

$$u_{2,1} + u_{2,-1} + (-1)^n(u_{2,1} - u_{2,-1}) - 2u_{1,1} - 2u = 0. \tag{4.26}$$

This is not in Kovalevskaya form, but, for each n, it can be solved for a $u_{2,J}$:

$$u_{2,-1} = u_{1,1} + u, \quad (n \text{ odd}), \qquad u_{2,1} = u_{1,1} + u, \quad (n \text{ even}).$$

A PΔE in Kovalevskaya form is then obtained by shifting each of these in n:

$$u_{2,0} = \pi_E(n)(u_{1,2} + u_{0,1}) + \pi_O(n)(u_{1,0} + u_{0,-1}), \tag{4.27}$$

where π_E and π_O are the even and odd projectors, respectively. ▲

The finiteness condition in Theorem 4.11 is important. Without it, one has to deal with singularities, where the solution of an IVP blows up. For some IVPs, singularities occur only when particular initial conditions are chosen (see Exercise 2.1(a)); we describe a set of initial conditions as *generic* if it produces no singularities[6]. However, it is also possible to construct IVPs that blow up for all initial conditions (see Exercise 2.1(b)). To avoid these complications, we will consider only IVPs with generic initial conditions.

The idea behind Theorem 4.11 may be generalized as follows. An IVP is *recursively computable* if it has a unique solution that can be calculated for each point in the domain in turn (for some ordering).

Example 4.13 The following PΔE is not generally solvable for a single $u_{1,j}$:

$$u_{1,0} - H(n-1)u_{1,-1} - H(-n-1)u_{1,1} = u_{0,1} - u, \qquad m \geq 0. \qquad (4.28)$$

Here $H(n)$ is the *unit step function* (or discrete Heaviside function):

$$H(n) = \begin{cases} 1, & n \geq 0; \\ 0, & n < 0. \end{cases} \qquad (4.29)$$

Nevertheless, the IVP can be solved for all initial conditions, $\{u(0,n) : n \in \mathbb{Z}\}$. The key is to start at $(0,0)$, where the PΔE yields $u(1,0) = u(0,1) - u(0,0)$. This acts as an initial condition for the line $m = 1$, where u is found from

$$u_{1,0} = \begin{cases} u_{1,-1} + u_{0,1} - u, & n \geq 1; \\ u_{1,1} + u_{0,1} - u, & n \leq -1. \end{cases}$$

The solution on each line of constant m is obtained by iterating this process (see Exercise 4.6). This example illustrates the importance of the ordering; for instance, it is essential to begin at $(m,n) = (0,0)$.   ▲

So far, we have considered only domains of the form $m \geq m_0$. As the above example suggests, one may also provide initial conditions on lines that are not parallel (for a suitably restricted domain), provided that the resulting IVP is recursively computable.

Example 4.14 The *discrete potential Korteweg–de Vries equation* (dpKdV),

$$(u_{1,1} - u)(u_{1,0} - u_{0,1}) = \beta(n) - \alpha(m) \neq 0, \qquad m \geq 0, \ n \geq 0, \qquad (4.30)$$

---

[6] It is interesting to know what happens as the initial conditions approach values for which the solution blows up. Typically, a near-singularity propagates as m increases. If the PΔE is a discrete integrable system, however, each near-singularity may occur at just a few points; in the limit, u stays finite at all other points. This remarkable property of *singularity confinement* is used as a (somewhat imperfect) test for integrability, in much the same way as the Painlevé test is used for PDEs. For an introduction to integrability tests, see Grammaticos et al. (2009).

is not in Kovalevskaya form. However, given a generic set of initial conditions, $\{u(m, 0) : m \in \mathbb{N}_0\}$ and $\{u(0, n) : n \in \mathbb{N}\}$, the IVP is recursively computable: at each step,

$$u_{1,1} = u + \frac{\beta(n) - \alpha(m)}{u_{1,0} - u_{0,1}}$$

yields each $u(m, n)$ in turn from what is already known.   ▲

It is worth identifying the main features of a recursively computable IVP, as illustrated by the last two examples.

- Each $(m, n) \in D$ is visited just once in the course of the computation.
- At the step when (m, n) is substituted into the PΔE, the value of u is known at every point of $\mathcal{S}(m, n)$ except one. That point is a vertex, $\mathcal{V}(m, n)$, of $\mathcal{S}(m, n)$; if the domain is regular then $\mathcal{V}(m, n) \in V_\mathcal{S}$.
- At each (m, n), the PΔE can be rearranged so that u at $\mathcal{V}(m, n)$ is determined uniquely from what is already known. In any coordinate system $(\hat{m}, \hat{n})$ in which $\mathcal{V}(m, n)$ is the rightmost point in $\mathcal{S}(m, n)$ and the leftmost vertex or border is at $\hat{m} = 0$, the PΔE is (locally) in Kovalevskaya form, with $\mathcal{V}(m, n)$ as the leading vertex. This ensures that the solution is single-valued.

These features define the circumstances under which a given PΔE can be solved recursively. The stencil places limits on which domains are allowable and the location of the necessary initial conditions. In particular, if $\mathcal{S}(m, n)$ has a rightmost border for every $(m, n) \in D$, there is no ordering for which the IVP on $m \geq m_0$ is recursively computable. For instance, the stencil for the dpKdV equation, at every point in the domain, is

$$\mathcal{S} = \{(0, 0), (1, 0), (0, 1), (1, 1)\}.$$

Initial conditions on $\{u(m_0, n) : n \in \mathbb{Z}\}$ alone will not give the solution on $m \geq m_0$, because u and $u_{0,1}$ are insufficient to determine both $u_{1,0}$ and $u_{1,1}$. By contrast, the initial conditions in Example 4.14 give enough information to determine $u_{1,1}$ everywhere in the upper-right quadrant.

4.5 Solutions that fill the plane

It is particularly useful to identify the circumstances under which a given PΔE can be solved recursively on the whole plane. To avoid ambiguity in the choice of ordering, we look only at problems for which $\mathbb{Z}^2$ is a regular domain and the PΔE can be written in Kovalevskaya form. More is needed, for (in general) Kovalevskaya form only allows the solution to be calculated to the right (that is, in the direction of increasing m). To cover the whole plane, the solution must also extend to the left. Accordingly, we describe a PΔE,

$$u_{p,J_1} = \omega(m, n, u_{0,J_2}, \mathbf{u}_{\{1,\cdot\}}, \dots, \mathbf{u}_{\{p-1,\cdot\}}), \tag{4.31}$$

as being in *bidirectional Kovalevskaya form* if it can also be written as

$$u_{0,J_2} = \Omega(m, n, \mathbf{u}_{\{1,\cdot\}}, \dots, \mathbf{u}_{\{p-1,\cdot\}}, u_{p,J_1}). \tag{4.32}$$

When the PΔE is written as (4.31), we call $(0, J_2)$ the *trailing vertex*. (When (4.32) is used to calculate the solution in the direction of decreasing m then $(0, J_2)$ is leading and (p, J_1) is trailing.) A PΔE in bidirectional Kovalevskaya form is analogous to an OΔE that can be written in both forward and backward form: provided that ω and Ω remain finite, the solution is obtained recursively from generic initial conditions on $m = m_0, \dots, m_0 + p - 1$, for any choice of m_0. The solution can be calculated, for one line of constant m at a time, from the values of u on the preceding p lines. It is helpful to regard these values as data on a fixed set of p initial lines in the (i, j)-plane. For instance, the PΔE (4.27) in Example 4.12 is in bidirectional Kovalevskaya form (with (4.26) giving the 'backward' equation). This PΔE is second-order in m, so the appropriate initial lines are $i = 0$ and $i = 1$.

When can a given PΔE be written in bidirectional Kovalevskaya form? The answer is determined by the set of vertices that can be leading vertices in Kovalevskaya form (in whichever coordinates are appropriate). We need to identify whether there exists a coordinate system in which a pair of these vertices act as the leading and trailing vertices.

Example 4.15   The dpKdV equation,

$$(u_{1,1} - u)(u_{1,0} - u_{0,1}) = \beta(n) - \alpha(m), \qquad m, n \in \mathbb{Z}, \tag{4.33}$$

has four vertices at which u can be isolated. The shear $(\hat{m}, \hat{n}) = (m + n, n)$ transforms this PΔE to bidirectional Kovalevskaya form:

$$\hat{u}_{2,1} = \hat{u} + \frac{\beta(\hat{n}) - \alpha(\hat{m} - \hat{n})}{\hat{u}_{1,0} - \hat{u}_{1,1}}, \qquad \hat{m}, \hat{n} \in \mathbb{Z}. \tag{4.34}$$

In these coordinates, the points $(2, 1)$ and $(0, 0)$ are the leading and trailing vertices, respectively. The initial lines are $\hat{\imath} = 0$ and $\hat{\imath} = 1$, that is, $j = -i$ and $j = 1 - i$, respectively. The points on these lines are sometimes described as a *staircase* (Figure 4.6).   ▲

Given a pair of vertices at which u can be isolated, one could use a variant of order minimization to determine whether or not there exists a lattice transformation that transforms the PΔE to bidirectional Kovalevskaya form. Stage I would be replaced by the condition that the $\hat{m}$-coordinate of the leading (resp.

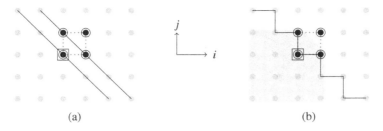

Figure 4.6 (a) Two initial lines are needed to solve (4.33) on $\mathbb{Z}^2$, with the vertices $(1, 1)$ leading and $(0, 0)$ trailing. (b) The initial data points, regarded as a staircase. The points in the shaded region belong to the shadow of $(1, 1)$.

trailing) vertex is greater (resp. less) than the $\hat{m}$-coordinate of every other vertex. This approach requires careful juggling of inequalities. Moreover, it is inefficient if one wishes to test all possible pairs of vertices, in order to find a coordinate system in which p is as small as possible.

There is an equivalent (but much neater) geometrical approach. We say that a point **i** is *reachable* from the vertex $\mathcal{V}$ if there exists a coordinate system $(\hat{m}, \hat{n})$ in which $\mathcal{V}$ is the leading vertex and $\hat{\imath}(\mathcal{V}) \leq \hat{\imath}(\mathbf{i})$. Given appropriate generic data on $\hat{\imath} < \hat{\imath}(\mathcal{V})$, one can use the P$\Delta$E to calculate $u_{\mathbf{i}}$. Some points are not reachable from $\mathcal{V}$; we call the set of all such points the *shadow* of $\mathcal{V}$ (see Figure 4.6). If (temporarily) one treats $\mathcal{V}$ as the origin, the shadow is the smallest sector of the plane that contains the stencil, excluding $\mathcal{V}$ itself. For any vertex at which u cannot be isolated, the shadow is the whole plane.

Given any pair of vertices, the intersection of the corresponding shadows is the set of points in the (i, j)-plane at which u is unreachable from either vertex. At each recursive step, one needs to know u at every unreachable stencil point and at the vertex that is trailing. Therefore, for a PΔE that is in bidirectional Kovalevskaya form in $(\hat{m}, \hat{n})$ coordinates, each of these points must lie on one of the p initial lines, $\hat{\imath} = 0, \ldots, p-1$.

To determine whether or not PΔE can be cast in bidirectional Kovalevskaya form and, if it can, to find the minimal set of initial lines, proceed as follows. Create a *shadow diagram* by translating each of the two vertices (with their shadows) to the origin. The problem has a solution if and only if the points that are not in either translated shadow include an infinite line, $\ell_0(\mathbf{k})$, through the origin. If the problem has a solution, there are infinitely many such lines; use the unique one that minimizes $\mathbf{k} \cdot \mathbf{k}$. Let $\mathcal{V}_1$ denote the vertex whose translated shadow is encountered first as the vector $\mathbf{k}$ is rotated anticlockwise about the origin in the shadow diagram, let $\mathcal{V}_2$ be the other vertex in the pair and let $\mathbf{l}$ be the vector from $\mathcal{V}_1$ to $\mathcal{V}_2$.

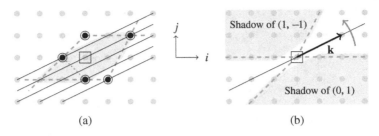

(a)   (b)

Figure 4.7 (a) To solve (4.35) on $\mathbb{Z}^2$, with leading and trailing vertices $(0, 1)$ and $(1, -1)$, five initial lines are needed. (b) The shadow diagram: $\mathbf{k}$ and $-\mathbf{k}$ are the closest points to the origin that lie on a shadow-free line.

Lemma 4.16 *Given generic initial conditions on $p = \det(\mathbf{k}\ \mathbf{l})$ consecutive lines that are parallel to $\ell_0(\mathbf{k})$, the values of u can be found at all points in $\mathbb{Z}^2$. Furthermore, p is the smallest number for which this is possible.*

The proof is a somewhat lengthy exercise in elementary coordinate geometry (Exercise 4.9). The method above is much simpler to use than it sounds, as the following example shows.

Example 4.17 We saw in Example 4.10 that the PΔE

$$u_{1,-1} + u_{0,1} + u_{-1,0} - (u_{2,1} - u_{0,-1})^2 = 0 \tag{4.35}$$

may be put into Kovalevskaya form with either $(0, 1)$ or $(1, -1)$ as the leading vertex. Now let us check whether these two vertices can be used to obtain u on the plane. Figure 4.7 shows how the optimal set of initial lines is determined. The stencil is shown in Figure 4.7(a); the finite set of points in the shaded region is the intersection of the two shadows. In the shadow diagram in Figure 4.7(b), there are lines through the origin that do not intersect either shadow. The one that we need has $\mathbf{k}$ as close to the origin as possible; the solution, up to an irrelevant choice of sign, is

$$\mathbf{k} = \begin{bmatrix} 2 \\ 1 \end{bmatrix}, \qquad \mathbf{l} = \begin{bmatrix} -1 \\ 2 \end{bmatrix},$$

which yields $p = 5$. The five initial lines are shown in Figure 4.7(a). Here p happens to coincide with the minimal order of the PΔE when $(0, 1)$ is the right-hand vertex. As the optimal $\mathbf{k}$ is determined by the shadows of both vertices, this sort of coincidence does not occur in general (see Exercise 4.10). ▲

In Example 4.17, the intersection of the two shadows in the (i, j)-plane is a finite set. It turns out that this is a necessary and sufficient condition for the existence of a coordinate system in which the corresponding vertices are leading and trailing, respectively (see Exercise 4.7).

The question of how many initial lines are needed is independent of most details of the PΔE. Two PΔEs that (up to a lattice transformation) have the same V_S, with the same pair of vertices at which u can be isolated, will require the same number of consecutive parallel initial lines.

So far, we have considered only PΔEs that can be written in bidirectional Kovalevskaya form. There is another common format for initial conditions, which is used when an initial line is a border. To cover $\mathbb{Z}^2$, at least one of the remaining initial lines must cross this border[7].

Example 4.18 For the dpKdV equation and other equations with the same stencil, it is common to use a cross of two initial lines, $i = 0$ and $j = 0$. For this to yield a solution on $\mathbb{Z}^2$, one must be able to isolate the value of u at each of the four vertices. By contrast, given the staircase of initial data shown in Figure 4.6, it is only necessary to isolate u and $u_{1,1}$. ▲

IVPs have a direct connection to symmetry methods. The symmetry condition for a given difference equation requires that the set of solutions is preserved. To write down the equation that determines the symmetries, one must be able to evaluate expressions on solutions. We have done this for a given OΔE in forward form, $u_p = \omega$, by substituting ω for each u_p. In effect, one writes the expression in terms of $u, \ldots, u_{p-1}$, which are regarded as free variables on a fixed set of initial points. (Such points are analogous to the initial lines that are needed for PΔEs with two independent variables.) Symmetries (and first integrals) are found by varying these initial variables independently.

For PΔEs, Kovalevskaya form is a natural generalization of forward form. It allows one to write expressions in terms of the values of u on the initial lines, and hence to find symmetries and conservation laws. In principle, this can be done whenever an IVP is recursively computable – in practice, the need to take into account details of the ordering may make this computationally expensive. From here on, therefore, we deal only with PΔEs that can be transformed to Kovalevskaya form, restricting attention to regular domains. However, we will choose whichever initial lines are most suitable for our purposes. For instance, the simplest way to calculate symmetries of the dpKdV equation (and many other equations with the same stencil) uses a cross of initial lines.

---

[7] A PΔE with initial conditions that lie on intersecting lines is known as a *Goursat problem* – see van der Kamp (2009).

Example 4.19   The autonomous dpKdV equation is

$$u_{1,1} = u + \frac{1}{u_{1,0} - u_{0,1}} . \tag{4.36}$$

(Any PΔE that does not depend explicitly on the independent variables is said to be *autonomous*.) On solutions of this equation, every $u_{i,j}$ can be written in terms of the values of u on the cross $ij = 0$. For example, replacing i by $i-1$ (and rearranging) gives

$$u_{-1,1} = u + \frac{1}{u_{-1,0} - u_{0,1}} .$$

Similarly,

$$u_{2,1} = u_{1,0} + \frac{1}{u_{2,0} - u_{1,1}} = u_{1,0} + \frac{u_{1,0} - u_{0,1}}{(u_{2,0} - u)(u_{1,0} - u_{0,1}) - 1} ,$$

and so on. To evaluate an expression (such as the linearized symmetry condition) on solutions, replace every $u_{i,j}$ for which $ij \neq 0$ by its equivalent in terms of the initial variables $u_{i,0}$ and $u_{0,j}$.   ▲

We now generalize the idea of a symmetry of a given difference equation,

$$\mathcal{A}(\mathbf{n}, [\mathbf{u}]) = 0, \qquad \mathbf{n} \in D. \tag{4.37}$$

Here $[\mathbf{u}]$ denotes $\mathbf{u}$ and a finite number of its shifts, $\mathbf{u_i}$, and the domain D is regular. Suppose that there exists a set, $\mathcal{I}$, of initial variables whose values determine the solution of the appropriate IVP uniquely. Suppose also that this solution depends smoothly on each $\mathbf{u_i} \in \mathcal{I}$. Then a *generalized symmetry* of (4.37) is a transformation of the form

$$\Psi : (\mathbf{n}, \mathcal{I}) \mapsto (\hat{\mathbf{n}}, \hat{\mathcal{I}}) = (\nu(\mathbf{n}), \psi_{\mathbf{n}}(\mathcal{I})), \tag{4.38}$$

where ν is a valid lattice map that preserves D^+ and where each $\psi_{\mathbf{n}}$ is a local diffeomorphism on the fibre of initial variables. For OΔEs of order p, this fibre is p-dimensional, whereas for PΔEs with an infinite solution domain, the fibre is infinite-dimensional. The transformation (4.38) is prolonged to the remaining variables in D^+ by using the PΔE (4.37). By construction, the set of solutions of the PΔE is preserved, so the transformation satisfies the *symmetry condition*,

$$\mathcal{A}(\hat{\mathbf{n}}, [\hat{\mathbf{u}}]) = 0 \qquad \text{when (4.37) holds.} \tag{4.39}$$

Just as for OΔEs, one can obtain Lie symmetries by linearizing the symmetry condition about the identity transformation (see Chapter 5).

4.6 Symmetries from valid lattice maps

4.6.1 Trivial symmetries

A *trivial symmetry* is a generalized symmetry, τ, that preserves every solution of the difference equation. For the graph, $\mathcal{G}_f$, of each solution to be preserved, the symmetry must be of the form

$$\tau : (\mathbf{n}, f(\mathbf{n})) \mapsto (\hat{\mathbf{n}}, f(\hat{\mathbf{n}})), \qquad \hat{\mathbf{n}} = \nu(\mathbf{n}), \qquad \mathbf{n} \in D,$$

where ν is a valid lattice map that preserves D^+. In other words, a trivial symmetry relabels the points in the graph in accordance with a valid change of lattice coordinates.

Consider a first-order OΔE that can be written in forward and backward form,

$$u_1 = \omega(n, u), \qquad u = \Omega(n, u_1), \qquad n \in \mathbb{Z}.$$

The translation $\nu_1 : n \mapsto n+1$ moves each point in the graph one step in the n-direction, so $\mathcal{G}_f$ is transformed as follows:

$$\nu_1 : (n, f(n)) \mapsto (n+1, f(n+1)) = (n+1, \omega(n, f(n))).$$

As this holds for each solution $u = f(n)$, the trivial symmetry corresponding to ν_1 is

$$\tau_{\nu_1} : (n, u) \mapsto (n+1, \omega(n, u)). \tag{4.40}$$

Similarly, $(\nu_1)^2$ yields the trivial symmetry

$$(\tau_{\nu_1})^2 : (n, u) \mapsto \left(n+2, \omega(n+1, \omega(n, u))\right),$$

and so on. The inverse of τ_{ν_1} is the trivial symmetry

$$\tau_{\nu_1^{-1}} : (n, u) \mapsto (n-1, \Omega(n-1, u)). \tag{4.41}$$

Moreover, the reflection $R : n \mapsto -n$ also yields a trivial symmetry, because whichever form of the OΔE is appropriate can be used to write $u(-n)$ in terms of n and $u(n)$. So every valid lattice transformation of $\mathbb{Z}$ gives rise to a trivial symmetry.

Example 4.20 The trivial symmetries of $u_1 = 2u+1-n$ are generated by

$$\tau_{\nu_1} : (n, u) \mapsto (n+1, 2u+1-n), \qquad \tau_R : (n, u) \mapsto (-n, 2^{-2n}(u-n)-n).$$

(To calculate τ_R easily, replace n by $-n$ in the first integral $\phi = 2^{-n}(u-n)$.) ▲

The same idea can be applied to OΔEs of order $p > 1$, whose initial values belong to the p-dimensional fibre with coordinates $(u, u_1, \ldots, u_{p-1})$.

Example 4.21   The trivial symmetries of $u_2 = 2u_1 - u$ are generated by

$$\tau_{\nu_1} : (n, u, u_1) \mapsto (n+1, u_1, 2u_1 - u),$$
$$\tau_R : (n, u, u_1) \mapsto (-n, (2n+1)u - 2nu_1, (2n+2)u - (2n+1)u_1).$$

(Again, the fact that first integrals are constant on solutions is useful.)   ▲

So far, we have considered only OΔEs whose domain is $\mathbb{Z}$. For a regular domain that is a finite discrete interval, $D = [n_0, n_1] \cap \mathbb{Z}$, the solution domain is $D^+ = [n_0, n_1 + p] \cap \mathbb{Z}$. Apart from the identity map, the only valid lattice map that preserves D^+ is the reflection $n \mapsto n_0 + n_1 + p - n$. Consequently, this reflection generates the group of trivial symmetries. If D is infinite, but is not $\mathbb{Z}$, the only valid lattice map that preserves D^+ is the identity map, so the OΔE has no other trivial symmetries.

Example 4.22   Consider the first-order OΔE

$$u_1 = \frac{u}{6 - n}, \qquad n = 0, 1, \ldots, 5,$$

whose general solution is $u = c_1(6 - n)!$ on $D^+ = \{0, 1, \ldots, 6\}$. The group of trivial symmetries is generated by the reflection

$$(n, u) \mapsto \left(6 - n, \frac{n! \, u}{(6 - n)!} \right), \qquad n = 0, 1, \ldots, 6.$$ ▲

The definition of a trivial symmetry applies equally to PΔEs; again, the graph of each solution, $\mathcal{G}_f = \{(\mathbf{n}, f(\mathbf{n})) : \mathbf{n} \in D^+\}$, must be preserved. For a scalar PΔE with two independent variables, the graph can be represented as a pinscreen (see Figure 4.8). The action of a trivial symmetry on the graph is easy to visualize: it maps the point at the top of each pin to another such point.

Example 4.23   The (discrete) *wave equation*,

$$u_{1,1} - u_{1,0} - u_{0,1} + u = 0, \qquad (m, n) \in \mathbb{Z}^2, \tag{4.42}$$

has the general solution

$$u(m, n) = u(m, 0) + u(0, n) - u(0, 0). \tag{4.43}$$

The reflection $v(m, n) = (n, m)$ produces the trivial symmetry

$$(m, n, u(m, n)) \mapsto (n, m, u(n, m)).$$

It is easy to show that the restriction of this transformation to the initial cross $mn = 0$ determines the transformed solution at all points (m, n).   ▲

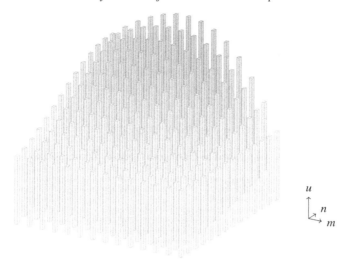

Figure 4.8 Pinscreen representation of a solution of a scalar PΔE with $M = 2$: the height of the pin at each (m, n) is the value $u = f(m, n)$.

Two generalized symmetries (of a given difference equation), Γ_1 and Γ_2, are said to be *equivalent* if $\Gamma_1^{-1} \circ \Gamma_2$ is a trivial symmetry. We know that every fibre-preserving transformation amounts to a composition of a lattice transformation and a transformation that does not change $\mathbf{n}$. Similarly, every symmetry, Γ, is equivalent to a symmetry, $\tilde{\Gamma}$, that acts only on the prolonged fibres[8], changing $\mathbf{u}$ (and its shifts) but not $\mathbf{n}$. We call $\tilde{\Gamma}$ the *fibre representative* of Γ.

Trivial symmetries do not change any solution, so they cannot be used to simplify or solve a given difference equation. (The term 'trivial' applies because the solutions, which are the fundamental objects of interest, are entirely unaffected.) Consequently, for most purposes, it is sufficient to consider only the fibre representatives of generalized symmetries.

4.6.2 Lattice symmetries

So far, we have looked at lattice transformations mainly from the passive viewpoint, seeking coordinates in which a given difference equation has a convenient form. From the active viewpoint, there are some lattice transformations that do not change the equation; we call these *lattice symmetries*. For the lattice transformation $(\hat{\mathbf{n}}, \hat{u}) = (\nu(\mathbf{n}), u)$ to be a lattice symmetry, ν must preserve

---

[8] In general, this decomposition of a fibre-preserving symmetry produces components that are defined only on a suitably prolonged space (see Exercise 4.13).

both the solution domain, D^+, and (up to a translation) the stencil, S. So it is only necessary to consider lattice transformations that satisfy these basic constraints. We do this from here on.

Let $(\hat{\mathbf{n}}, \hat{u}) = (\nu(\mathbf{n}), u)$ be a lattice transformation that preserves D^+ and S, up to a translation of the stencil. Let $u = f(\mathbf{n})$ be a solution of the difference equation; its graph is

$$\mathcal{G}_f = \{(\mathbf{n}, f(\mathbf{n})) : \mathbf{n} \in D^+\}.$$

Each point, $(\mathbf{n}, f(\mathbf{n}))$, is mapped to $(\nu(\mathbf{n}), f(\mathbf{n}))$, so $\mathcal{G}_f$ is mapped to

$$\mathcal{G}_{\tilde{f}} = \{(\mathbf{n}, \tilde{f}(\mathbf{n})) : \mathbf{n} \in D^+\},$$

where

$$\tilde{f}(\mathbf{n}) = f\left(\nu^{-1}(\mathbf{n})\right). \tag{4.44}$$

The lattice transformation is a lattice symmetry of the difference equation if $\mathcal{G}_{\tilde{f}}$ is the graph of a solution whenever $\mathcal{G}_f$ is the graph of a solution.

For each lattice symmetry, the fibre representative satisfies

$$\tilde{\nu} : (\mathbf{n}, f(\mathbf{n})) \mapsto \left(\mathbf{n}, \tilde{f}(\mathbf{n})\right), \qquad \mathbf{n} \in D^+,$$

for every solution, $u = f(\mathbf{n})$. Just as for trivial symmetries, the fibre representative of a lattice symmetry can be written in terms of an appropriate set of initial variables.

Example 4.24   As a simple example, consider the OΔE

$$u_2 + \alpha\, u_1 + \beta\, u = 0, \qquad n \in \mathbb{Z}. \tag{4.45}$$

The translation $(n, u) \mapsto (n + 1, u)$ maps $D^+ = \mathbb{Z}$ to itself. The fibre representative is $(n, u) \mapsto (n, u_1)$; this maps the OΔE to

$$u_3 + \alpha\, u_2 + \beta\, u_1 = 0, \qquad n \in \mathbb{Z},$$

which amounts to (4.45). Similarly, translation by any $k \in \mathbb{Z}$ is a lattice symmetry, so the homogeneous OΔE (4.45) has a subgroup of lattice symmetries that is isomorphic to the additive group $\mathbb{Z}$ under addition. This result is easily generalized to all difference equations that are preserved by translations of the independent variables (see Exercise 4.15).

The fibre representative of the reflection $(n, u) \mapsto (-n, u)$ transforms $u(n+i)$ to $u(-n-i)$; the transformed OΔE is

$$u + \alpha\, u_1 + \beta\, u_2 = 0, \qquad n \in \mathbb{Z},$$

so the reflection is a lattice symmetry if and only if $\beta = 1$.   ▲

So far, we have considered only rigid lattice maps. One might think that rigidity is necessary to preserve the stencil. However, there are some PΔEs with lattice symmetries that stem from non-rigid lattice maps.

Example 4.25   The lattice symmetries of

$$u_{0,1} = u_{1,0} + u, \qquad m \in \mathbb{Z}, \ n \geq 0, \tag{4.46}$$

include

$$v : (m, n, u) \mapsto (-m-n, n, u),$$

which is the composition of a unit shear in the m-direction and a reflection in $m = 0$.   ▲

Lattice symmetries are not the only kind of discrete symmetry. Many PΔEs have symmetries of the form

$$(\mathbf{n}, u(\mathbf{n})) \mapsto \left(\mathbf{n}, \ h(\mathbf{n}) u(v^{-1}(\mathbf{n}))\right), \qquad \mathbf{n} \in D^+, \tag{4.47}$$

where v is a valid lattice map that preserves D^+ and h is a given function.

Example 4.26   The autonomous dpKdV equation,

$$(u_{1,1} - u)(u_{1,0} - u_{0,1}) = 1, \qquad (m, n) \in \mathbb{Z}^2, \tag{4.48}$$

has lattice symmetries that stem from translations in the m- and n-directions. (These constitute an abelian group that is isomorphic to $\mathbb{Z}^2$ under addition.) The stencil is a square of points, so it is natural to ask whether (4.48) also has symmetries that stem from the dihedral group D_4, which preserves the stencil. A simple calculation shows that it does, provided that $h(m, n) = (-1)^{m+n}$ in (4.47). Specifically, the discrete symmetries of (4.48) include

$$(m, \ n, \ u(m, n)) \mapsto \left(m, \ n, \ (-1)^{m+n} u(n, -m)\right), \qquad \text{(rotation by } \pi/2),$$

$$(m, \ n, \ u(m, n)) \mapsto \left(m, \ n, \ (-1)^{m+n} u(n, m)\right), \qquad \text{(reflection)}. \qquad ▲$$

Notes and further reading

The lattice $\mathbb{Z}^N$ is used in various branches of pure and applied mathematics. For an accessible elementary introduction to its geometric properties, with applications to number theory, see Hardy and Wright (1979). Many of the concepts within this chapter are specific to difference equations. However, Kovalevskaya form, generalized symmetries, trivial symmetries and equivalent symmetries all have continuous counterparts. (For differential equations, the analogue of the fibre representative is known as the evolutionary form of a symmetry.) Olver (1993) describes these ideas in detail and deals carefully with the technical issues.

For differential equations, some of the main results on existence and uniqueness of solutions were discovered well over a century ago. By contrast, the corresponding theory for difference equations is still relatively incomplete. For a comprehensive overview of the Cauchy problem for difference equations, see Gregor (1998); this paper describes recursively computable IVPs in detail and proves some key results about the existence of solutions and the choice of a correct ordering. The geometric (stencil-based) approach to Cauchy and Goursat IVPs on $\mathbb{Z}^2$ was introduced by van der Kamp (2009).

The dpKdV equation is one of the simplest nonlinear PΔEs that is integrable in the usual sense – it has a (discrete) Lax pair, infinite hierarchies of symmetries and conservation laws, an auto-Bäcklund transformation, and various other features that are common to discrete integrable systems. Grammaticos et al. (2004) is a collection of lecture notes that surveys these ideas in the discrete setting. For a fuller description of lattice geometry, see Bobenko and Suris (2008); the geometric approach provides powerful tools for classifying discrete integrable systems (see Adler et al., 2003, 2009, 2012) and generating Lax pairs (see Bridgman et al., 2013). Many such systems were first discovered as discretizations of continuous integrable systems. Two very successful approaches were pioneered by Hirota (1977) and Quispel et al. (1984). Most techniques for dealing with integrable PΔEs are designed to exploit integrability. We will not discuss these further, as the purpose of this book is to describe methods that work whether or not a given difference equation happens to be integrable.

Exercises

4.1   Prove that the set of all valid lattice maps is a group, and show that the set of rigid lattice maps is a subgroup of this group.

4.2   Suppose that $\{\mathbf{m}_1, \mathbf{m}_2\}$ is a basis for $\mathbb{Z}^2$ and that a 2×2 matrix A satisfies

$$A\mathbf{m}_1 = \mathbf{m}_1, \qquad A\mathbf{m}_2 = \mathbf{m}_2 + r\mathbf{m}_1, \quad r \in \mathbb{Z}.$$

Show that $\det(A) = 1$ and find the condition that r must satisfy for A to be the matrix of a valid lattice map. For the smallest positive r that satisfies this condition, write A in terms of the matrices S and T, which are defined in (4.7).

4.3   Prove that for every border in a stencil, there exists a lattice transformation that makes the border rightmost. To achieve this, find a transformation such that the transformed border is a line $\hat{m} = m_0$ and every point in the transformed stencil satisfies $\hat{m} \leq m_0$. [Hint: the Euclidean algorithm may be helpful.]

4.4 Find a lattice transformation that minimizes the orders of the PΔE (4.17) as far as possible.

4.5 Calculate the general solution of the PΔE

$$u_{0,2} + (m+\alpha n)\, u_{1,1} - (m+\alpha n + 1)\, u_{2,0} = 0,$$

when (a) $\alpha = 1$; (b) $\alpha = 0$.

4.6 (a) Calculate the solution of the IVP in Example 4.13 with initial conditions $u(0, n) = 1/(1 + n^2)$, $n \in \mathbb{Z}$. [Try to solve the problem for arbitrary initial conditions first, treating the regions $n \geq 0$ and $n < 0$ separately.]

(b) Show that (4.28) cannot be put into bidirectional Kovalevskaya form by any lattice transformation.

(c) Find a set of initial points at which the values of u determine a unique solution of (4.28) on $\mathbb{Z}^2$.

4.7 (a) Sketch the shadow diagram for the dpKdV equation with $(1, 0)$ and $(0, 1)$ as the leading and trailing vertices, respectively. Find the minimum number of parallel initial lines that are required, together with their direction.

(b) Now try to repeat the above with $(0, 1)$ replaced by $(1, 1)$. What goes wrong? How does this observation generalize to every pair of vertices that share a border?

4.8 (a) Find a lattice transformation that minimizes the orders of

$$u_{3,0} - u_{-2,2} = \sin u, \tag{4.49}$$

subject to the transformed PΔE being in Kovalevskaya form.

(b) What is the least number of parallel initial lines that will yield a solution of (4.49) on $\mathbb{Z}^2$?

4.9 This exercise compiles the results that lead to the proof of Lemma 4.16.

(a) Prove that for any two vectors $\mathbf{a}, \mathbf{b} \in \mathbb{Z}^2$, premultiplying each vector by any $A \in \mathrm{SL}_2(\mathbb{Z})$ does not alter $\det(\mathbf{a}\ \mathbf{b})$. Use this to conclude that, in the notation of §4.1.1, $\det(\hat{\mathbf{e}}_1\ \hat{\mathbf{e}}_2) = 1$ whenever $\det(A) = 1$.

(b) Use the above results to determine the spacing between consecutive lines that are parallel to $\ell_0(\hat{\mathbf{e}}_1)$. Hence show that any vector $\mathbf{a} \in \mathbb{Z}^2$ which does not lie on such a line is straddled by $|\det(\hat{\mathbf{e}}_1\ \mathbf{a})| - 1$ of them (not counting lines at the endpoints of $\mathbf{a}$).

(c) Explain why it is necessary that there exists a line in the shadow diagram that does not intersect either shadow. [Hint: what happens to the chosen pair of vertices in a coordinate system in which this condition is not met?]

(d) Now complete the final details of the proof. Specifically, how can p be minimized? Why is the minimal solution unique? Why is $\det(\mathbf{k}\ \mathbf{l})$ positive?

4.10 (a) Find the lattice transformation that puts (4.23) into Kovalevskaya form, with minimal order in $\hat{m}$ subject to $(0, 1)$ becoming the rightmost vertex.

(b) Repeat (a), but with $(-1, 0)$ becoming the rightmost vertex.

(c) Now determine the minimal number of parallel initial lines that will enable (4.23) to be solved on $\mathbb{Z}^2$, using $(0, 1)$ and $(-1, 0)$ as the leading and trailing vertices, respectively. Sketch your solution. What is the gradient of these lines?

(d) The following linear PΔE has the stencil (4.24):

$$u_{1,-1} + u_{0,1} + u_{-1,0} + u_{2,1} + u_{0,-1} = 0.$$

Find each pair of vertices that can be used as leading and trailing vertices in bidirectional Kovalevskaya form. Which of these pairs gives the lowest-order transformed PΔE?

4.11 Determine all trivial symmetries of

$$u_1 = \frac{u}{u - 1}, \qquad n \in \mathbb{Z}.$$

4.12 Find all lattice symmetries of the PΔE (4.46).

4.13 Show that every OΔE of the form

$$u_2 + \alpha\, u_1 + \beta u = 0, \qquad \alpha\beta \neq 0, \qquad n \in \mathbb{Z},$$

has the (fibre-preserving) symmetry $(n, u) \mapsto (n+1, 2u)$.

4.14 (a) Show that all solutions of the autonomous dpKdV equation (4.36) satisfy the following discrete *Toda-type equation*:

$$\frac{1}{u_{1,1} - u} - \frac{1}{u_{1,-1} - u} - \frac{1}{u_{-1,1} - u} + \frac{1}{u_{-1,-1} - u} = 0. \quad (4.50)$$

(b) Show that the transformation

$$(m, n, u) \mapsto (m + \pi_E(m+n), \; n + \pi_E(m+n), \; u) \quad (4.51)$$

(suitably prolonged) is a symmetry of (4.50). Here π_E is the even projector.

(c) Show that the transformation (4.51) is not a lattice symmetry of (4.50), even though it is a symmetry. Explain how this (seeming) paradox is resolved.

4.15 (a) Show that the PΔE

$$u_{2,1} + (m - n)\, u_{1,0} - 1/u = 0, \qquad (m, n) \in \mathbb{Z}^2,$$

has the lattice symmetry whose fibre representative is

$$(m, \; n, \; u(m, n)) \mapsto (m, \; n, \; u(m + 1, n + 1)).$$

(b) Find the conditions under which a PΔE is preserved by all translations along a given line, $\ell_0(\mathbf{k})$, and determine the fibre representatives of the corresponding lattice symmetries.

5

Solution methods for PΔEs

'... that's the same thing, you know.'
'Not the same thing a bit!' said the Hatter.

(Lewis Carroll, Alice's Adventures in Wonderland)

Relatively few PΔEs can be solved exactly. However, there are many techniques for obtaining *some* exact solutions of a given PΔE, typically by solving an associated OΔE. Generally speaking, these are adaptations of PDE methods. This chapter is a brief introduction to some of the most effective approaches.

5.1 Notation for PΔEs

We begin by describing index notation for PΔEs, which will enable us to state and prove some general results concisely. From here on, if the (regular) domain is not specified, it is assumed to be $\mathbb{Z}^N$. Suppose that a given system of PΔEs has N independent variables, $\mathbf{n} = (n^1, \ldots, n^N) \in \mathbb{Z}^N$, and M dependent variables, $\mathbf{u} = (u^1, \ldots, u^M)$. For most purposes, it is best to use lattice coordinates n^i that minimize the orders of the PΔE(s) as far as possible.

Having fixed the lattice coordinates, we will use $u_{\mathbf{J}}^\alpha$ to denote $u^\alpha(\mathbf{n} + \mathbf{J})$, for each multi-index $\mathbf{J} = (j^1, \ldots, j^N)$. For convenience, we write $\mathbf{J} = j^k \mathbf{1}_k$, where $\mathbf{1}_k$ is the multi-index whose k^{th} entry is 1 and whose other entries are zero. The forward shift operator in the kth direction is S_k, which is defined by

$$S_k : (\mathbf{n}, f(\mathbf{n})) \mapsto (\mathbf{n} + \mathbf{1}_k, f(\mathbf{n} + \mathbf{1}_k)). \tag{5.1}$$

Consequently,

$$S_k : \mathbf{u} \mapsto \mathbf{u}_{\mathbf{1}_k}, \qquad S_k : \mathbf{u}_{\mathbf{J}} \mapsto \mathbf{u}_{\mathbf{J}+\mathbf{1}_k}. \tag{5.2}$$

It is also useful to use the following shorthand:

$$S_{\mathbf{J}} = S_1^{j_1} S_2^{j_2} \cdots S_N^{j_N}. \tag{5.3}$$

The identity operator, which maps $\mathbf{n}$ to itself, is $I = S_{\mathbf{0}}$. A *linear difference operator* is an operator of the form

$$\mathcal{D} = a^{\mathbf{J}}(\mathbf{n}) \, S_{\mathbf{J}} \tag{5.4}$$

(summed over $\mathbf{J}$) that has at least two terms.

Although index notation is convenient for general results, it can complicate simple problems, particularly if it conflicts with well-established notation. Therefore, to understand how techniques are used in practice, we will usually consider only scalar P∆Es with two independent variables, provided that the generalization to higher dimensions is obvious. With this restriction, the general linear difference operator is

$$\mathcal{D} = a^{0,0}(m, n) \, \mathrm{I} \; + \sum_{\substack{i,j \in \mathbb{Z} \\ (i,j) \neq (0,0)}} a^{i,j}(m, n) \, S_m^i S_n^j.$$

(For clarity, the shift operators are written as S_m and S_n, rather than S_1 and S_2.) A scalar P∆E with two independent variables is linear if it is of the form

$$\mathcal{D}(u) = b(m, n),$$

that is, if each term contains at most one $u_{i,j}$. This P∆E is homogeneous (in u) if $b(m, n) = 0$.

5.2 Factorizable P∆Es

Suppose that a given scalar P∆E (which need not be linear) can be written as

$$\mathcal{D}(v) = 0, \tag{5.5}$$

where $\mathcal{D}$ is a linear difference operator and v depends on m, n and a finite number of shifts of u. Then the P∆E is equivalent to a lower-order equation,

$$v = \kappa(m, n), \tag{5.6}$$

where $\kappa(m, n)$ is an arbitrary element of the kernel of $\mathcal{D}$. The simplest case is when $\mathcal{D}$ is first-order in at least one of the variables. This approach has two advantages. Sometimes, one can factorize a higher-order $\mathcal{D}$ and solve (5.5) iteratively. Even if it is not possible to solve (5.6) for all $\kappa(m, n) \in \ker(\mathcal{D})$, one may obtain some solutions of (5.5) by restricting κ further.

Example 5.1 The discrete wave equation,

$$u_{1,1} - u_{1,0} - u_{0,1} + u = 0, \tag{5.7}$$

is a linear PΔE that can be written as

$$(S_m - I)(u_{0,1} - u) = 0.$$

Therefore

$$u_{0,1} - u = A(n),$$

for an arbitrary function A. This has a particular solution, $u = g(n)$, where $g(n) = \sigma_k\{A(k); 0, n\}$. Consequently, the general solution of (5.7) is

$$u = f(m) + g(n),$$

where f and g are arbitrary functions. ▲

Example 5.2 Find the general solution of the nonlinear PΔE

$$u_{1,1}u_{1,0} - 2u_{0,1}u = 0.$$

Solution: The PΔE has an obvious linear factor, as follows:

$$(S_m - 2I)(u_{0,1}u) = 0.$$

Inverting the linear operator $S_m - 2I$, we obtain

$$u_{0,1}u = 2^m A(n),$$

where A is arbitrary. This is a Riccati equation in which m is treated as a parameter. By using the methods of Chapter 1, we obtain the general solution,

$$u = 2^{m/2}g(n)\left[f(m)\right]^{(-1)^n},$$

where f, g are arbitrary and $f(m)$ is nonzero for all $m \in \mathbb{Z}$. ▲

In the last two examples, $\mathcal{D}$ is, in effect, an ordinary difference operator in a particular direction. More generally, to find the kernel of a given linear operator that involves shifts in each direction, one must be able to solve the linear PΔE

$$\mathcal{D}(u) = 0. \tag{5.8}$$

Suppose that $\mathcal{D}$ is first-order in each direction, so that (5.8) is of the form

$$(S_m - a(m,n)S_n - b(m,n)I)\, u = 0. \tag{5.9}$$

If the coefficients a, b are nonzero constants, (5.9) can be written as

$$u = (aS_n + bI)S_m^{-1}u. \tag{5.10}$$

So the general solution of (5.10) on $m \geq m_0$ is obtained by iteration:

$$u = (aS_n + bI)^{m-m_0} A(n), \tag{5.11}$$

where $A(n) = u(m_0, n)$ is arbitrary. The binomial expansion is then used to write u explicitly:

$$u = a^{m-m_0} \sum_{i=0}^{m-m_0} \binom{m - m_0}{i} \left(\frac{b}{a}\right)^{m-m_0-i} A(n + i). \qquad (5.12)$$

Similarly, if a and b are functions of n only, (5.11) is again the general solution of the PΔE (5.9). In this case, the binomial expansion (5.12) cannot be used, because (5.11) involves shifts of n in the functions a and b. Finally, if the coefficients depend upon both m and n, (5.10) is replaced by

$$u = \left\{S_m^{-1}(aS_n + bI)\right\} S_m^{-1} u.$$

Iteration yields the general solution

$$u = \left\{S_m^{-1}(aS_n + bI)\right\}\left\{S_m^{-2}(aS_n + bI)\right\} \cdots \left\{S_m^{m_0-m}(aS_n + bI)\right\} A(n). \qquad (5.13)$$

Example 5.3 To solve the linear homogeneous PΔE

$$u_{1,0} + u_{0,1} - (m + n + 1)u = 0, \qquad m \geq 0, \ n \geq 0,$$

set $m_0 = 0$ in (5.13). This gives

$$u = \{- S_n + (m + n)I\}\{- S_n + (m + n - 1)I\} \cdots \{- S_n + (n + 1)I\} A(n)$$

$$= \sum_{r=0}^{m} m^{(r)}(m + n)^{(r)}(-1)^{m-r} A(m + n - r)/r!,$$

where $k^{(r)}$ is defined by (1.14) and $A(n) = u(0, n)$. ▲

This approach works equally well if the operator $\mathcal{D}$ is first-order in just one direction. Suppose that (5.8) is of the form

$$\left(S_m - \widetilde{\mathcal{D}}\right) u = 0, \qquad (5.14)$$

where $\widetilde{\mathcal{D}}$ is an operator whose only shifts are in the variable n. Then (5.13) is replaced by

$$u = \left(S_m^{-1}\widetilde{\mathcal{D}}\right)\left(S_m^{-2}\widetilde{\mathcal{D}}\right) \cdots \left(S_m^{m_0-m}\widetilde{\mathcal{D}}\right) A(n). \qquad (5.15)$$

If $\widetilde{\mathcal{D}}$ is independent of m, this reduces to

$$u = \widetilde{\mathcal{D}}^{m-m_0}(A(n)). \qquad (5.16)$$

There are two difficulties with the above approach. First, (5.15) becomes increasingly long as m increases, unless it can be expressed in closed form. The second difficulty is that, although m_0 may be chosen arbitrarily, the above solutions are valid only for $m \geq m_0$. To find the solution for $m < m_0$, one must invert the operator $\widetilde{\mathcal{D}}$, writing u in terms of $u_{1,j}$ (for various j). This produces nested sums (of an arbitrary function) that cannot be simplified.

So far, we have focused on operators that are first-order in m; the same process could equally well be applied to operators that are first-order in n. Furthermore, it is straightforward (at least, in principle) to iterate the process, just as we did for the wave equation. If a PΔE has repeated factors, these can be dealt with simultaneously. For instance, if $\widetilde{\mathcal{D}}$ is independent of m, the general solution of

$$\left(\mathrm{S}_m - \widetilde{\mathcal{D}}\right)^k u = 0, \qquad k \geq 2, \tag{5.17}$$

is

$$u = m^{k-1}\widetilde{\mathcal{D}}^{m-m_0}(A_1(n)) + m^{k-2}\widetilde{\mathcal{D}}^{m-m_0}(A_2(n)) + \cdots + \widetilde{\mathcal{D}}^{m-m_0}(A_k(n)). \tag{5.18}$$

Here $A_1, \ldots, A_k$ are independent arbitrary functions.

Example 5.4 Solve the linear PΔE

$$u_{2,0} + u_{0,2} + u - 2(u_{1,1} + u_{1,0} - u_{0,1}) = 0, \qquad m \geq 0, \, n \in \mathbb{Z},$$

subject to $u(0, n) = n$ and $u(1, n) = 2n + 1 - 2^{-n}$.

Solution: The PΔE factorizes as follows:

$$(\mathrm{S}_m - \mathrm{S}_n - \mathrm{I})^2 u = 0.$$

From (5.11) and (5.18), the general solution of the PΔE is

$$u = m(\mathrm{S}_n + \mathrm{I})^m A_1(n) + (\mathrm{S}_n + \mathrm{I})^m A_2(n).$$

Applying the initial conditions in turn, we obtain

$$A_2(n) = u(0, n) = n, \qquad (\mathrm{S}_n + \mathrm{I})A_1(n) = u(1, n) - (\mathrm{S}_n + \mathrm{I})A_2(n) = -2^{-n}.$$

Therefore the solution of the initial-value problem is

$$u = (\mathrm{S}_n + \mathrm{I})^m(n) - m(\mathrm{S}_n + \mathrm{I})^{m-1}(2^{-n}).$$

It is a simple exercise to evaluate this solution in closed form:

$$u = 2^{m-1}(m + 2n) - 2^{1-m-n}3^{m-1}m. \qquad \blacktriangle$$

The methods described above produce the general solution of linear (and some nonlinear) factorizable PΔEs with first-order factors. However, relatively few PΔEs have a general solution; usually, one must restrict attention to solution that satisfy given initial conditions or boundary conditions.

5.3 Separation of variables

For linear PΔEs, there are methods that produce large families of solutions by exploiting the principle of linear superposition. Consider the class of autonomous linear homogeneous PΔEs; each PΔE in this class is of the form

$$P(S_m, S_n) u = 0, \tag{5.19}$$

where P is a polynomial with constant coefficients. Equation (5.19) admits solutions of the form

$$u = \mu^m v^n,$$

where $\mu, v \in \mathbb{C}$, whenever

$$P(\mu, v) = 0. \tag{5.20}$$

If $\mu = f(v)$ is a solution of (5.20) then

$$u = (f(v))^m v^n$$

is a solution for each v in the domain of f. By linear superposition, so is

$$u = \int A(v)(f(v))^m v^n \, dv, \tag{5.21}$$

where A is an arbitrary function (or generalized function) that is constrained only by the need for the integral to converge for each (m, n). If (5.20) has several solutions, $\mu = f_i(v)$, each of these yields a family of solutions (5.21). Therefore

$$u = \sum_i \int A_i(v)(f_i(v))^m v^n \, dv \tag{5.22}$$

is a solution for every choice of the arbitrary functions $A_i(v)$. Additionally, if (5.20) has solutions $v = c_j$, the corresponding solution of (5.19) is

$$u = \sum_j B_j(m)(c_j)^n, \tag{5.23}$$

where each B_j is an arbitrary function. Finally, the solutions (5.22) and (5.23) may be superposed. The above method is called *Lagrange's method*.

Example 5.5 Apply Lagrange's method to

$$u_{1,1} + u_{1,0} - u_{0,3} - u_{0,2} = 0. \tag{5.24}$$

This PΔE can be written in operator notation as

$$(S_n + I)\left(S_m - S_n^2\right) u = 0.$$

Therefore $P(\mu, \nu) = 0$ has two solutions: either $\mu = \nu^2$ or $\nu = -1$.

Consequently, the most general solution that can be obtained by Lagrange's method is

$$u = \int A(\nu)\nu^{2m+n}\,d\nu + B(m)(-1)^n. \tag{5.25}$$

The PΔE (5.24) has not yet been optimized for order; the transformation

$$\hat{m} = m, \quad \hat{n} = 2m + n, \quad \hat{u} = u$$

reduces (5.24) to the following (after shifting $\hat{n}$):

$$\hat{u}_{1,1} + \hat{u}_{1,0} - \hat{u}_{0,1} - \hat{u} = 0.$$

This has an easy factorization,

$$(S_{\hat{m}} - I)(S_{\hat{n}} + I)\hat{u} = 0,$$

which yields the general solution

$$\hat{u} = f(\hat{n}) + g(\hat{m})(-1)^{\hat{n}},$$

where f and g are arbitrary. Thus every solution of (5.24) is of the form

$$u = f(2m + n) + g(m)(-1)^n. \qquad \blacktriangle$$

This solution is suggested by the fact that $A(\nu)$ is arbitrary in (5.25), subject to the constraints that the integral converges and u is real-valued.

The method of *separation of variables* is a generalization of Lagrange's method that does not require the linear homogeneous PΔE to be autonomous. The method is to seek nonzero solutions that are products of functions of a single independent variable:

$$u(m, n) = v(m)w(n). \tag{5.26}$$

Rewrite the PΔE using this anzatz and divide by $v(m)w(n)$. If the result can be rearranged in the form

$$\text{function of } m = \text{function of } n,$$

both sides must equal a separation constant, λ (by analogy with separation of variables for PDEs). This gives a pair of OΔEs for v and w. The corresponding solution of the PΔE will be of the form

$$u = c(\lambda)v(m; \lambda)w(n; \lambda),$$

where $c(\lambda)$ is an arbitrary constant. Now combine all such solutions by linear superposition, to get

$$u = \int c(\lambda) v(m; \lambda) w(n; \lambda) \, d\lambda. \qquad (5.27)$$

Here, the integral is taken over all values of $\lambda \in \mathbb{C}$ such $vw \neq 0$. The constants $c(\lambda)$ are arbitrary, subject to the requirements that the integral converges and u is real-valued.

Example 5.6 For the non-autonomous P∆E

$$u_{1,0} - (n + 1) u_{0,1} + u = 0, \qquad m \in \mathbb{Z}, \ n \geq 0, \qquad (5.28)$$

the nonzero separable solutions of the form (5.26) satisfy

$$\frac{v_1}{v} = \lambda = \frac{(n + 1) w_1}{w} - 1, \qquad \lambda \notin \{-1, 0\}.$$

Up to arbitrary constant factors, the solution of this pair of O∆Es is

$$v(m; \lambda) = \lambda^m, \qquad w(n; \lambda) = \frac{(\lambda + 1)^n}{n!}.$$

Therefore the solution (5.27) of the P∆E (5.28) is

$$u = \frac{1}{n!} \int c(\lambda) \lambda^m (\lambda + 1)^n \, d\lambda = \sum_{k=0}^{n} \frac{1}{k! \, (n - k)!} \int c(\lambda) \lambda^{m+k} \, d\lambda.$$

This suggests that

$$u = \sum_{k=0}^{n} \frac{f(m + k)}{k! \, (n - k)!}$$

is a solution of (5.28) for arbitrary f, a result that is easy to verify. ▲

The method of separation of variables is particularly useful when initial or boundary conditions are given on lines of constant m or n (see Exercise 5.3).

5.4 Wavelike solutions of P∆Es

If a PDE for $u(x, t)$ is autonomous (that is, it has no explicit dependence on the independent variables x and t), one can seek travelling wave solutions of the form $u = f(x - ct)$, where c is an arbitrary constant. This ansatz reduces the PDE to an ODE, which may be solvable. Wavelike solutions of autonomous P∆Es can be found by a similar approach. For simplicity, we shall consider only P∆Es of the form

$$u_{1,1} = \omega(u, u_{1,0}, u_{0,1}), \qquad (5.29)$$

although the idea is equally applicable to higher-order PΔEs. The simplest wavelike solutions are those for which u is a function of m only or n only. In Example 5.1, we saw that every solution of the wave equation (5.7) is a linear combination of solutions of these forms. For nonlinear PΔEs, there is no direct superposition principle[1]; in general, one cannot superpose wavelike solutions. Nevertheless, sometimes one can obtain large families of wavelike solutions by using a simple ansatz.

Example 5.7   The nonlinear PΔE

$$u_{1,1} = -1/u_{1,0} - 1/u_{0,1} - u, \qquad (m,n) \in \mathbb{Z}^2, \qquad (5.30)$$

has solutions of the form $u = f(m)$, where

$$(S_m + I)(f(m) + 1/f(m)) = 0.$$

The general solution of this OΔE splits into two families:

$$f(m) = \epsilon(m)\sinh(c_1) + (-1)^m \cosh(c_1),$$
$$f(m) = \epsilon(m)\sinh(c_1) - (-1)^m \cosh(c_1).$$

Here c_1 is an arbitrary non-negative constant and ϵ is an arbitrary map from $\mathbb{Z}$ to $\{-1, 1\}$. As (5.30) is preserved by the exchange of m and n, there are also solutions of the form $u = f(n)$. Although these solutions cannot be linearly superposed in general, there is a two-parameter family of solutions

$$u = \tilde{c}_1(-1)^m + \tilde{c}_2(-1)^n. \qquad \blacktriangle$$

Many PΔEs have wavelike solutions that are not aligned with the coordinate axes. Such solutions may be found by using a valid lattice transformation to introduce new coordinates $(\tilde{m}, \tilde{n}) = \nu(m, n)$, then using the ansatz $u = f(\tilde{m})$.

Example 5.8   For the PΔE (5.30), the above ansatz applied to the coordinates $(\tilde{m}, \tilde{n}) = (m + kn, n)$ yields

$$f(\tilde{m} + k + 1) = -1/f(\tilde{m} + 1) - 1/f(\tilde{m} + k) - f(\tilde{m}).$$

In the previous example, we solved this OΔE with $k = 0$. When $k = 1$, the OΔE is

$$f(\tilde{m} + 2) = -2/f(\tilde{m} + 1) - f(\tilde{m}),$$

---

[1]   However, if a nonlinear PΔE is obtainable from a linear system, it can inherit the underlying superposition principle.

which is also quite easy to solve:

$$
f(\tilde{m}) = \begin{cases} c_2 \left(\dfrac{1 + c_1}{1 - c_1} \right)^{\tilde{m}/2}, & \tilde{m} \text{ even;} \\[3ex] \dfrac{c_1 - 1}{c_2} \left(\dfrac{1 - c_1}{1 + c_1} \right)^{(\tilde{m}-1)/2}, & \tilde{m} \text{ odd.} \end{cases}
$$

Here $c_2 \neq 0$ and $(c_1)^2 \neq 1$. The corresponding solution of (5.30) is obtained by replacing $\tilde{m}$ by $m + n$.   ▲

Example 5.9 Find all wavelike solutions of

$$
u_{1,1} = u(\lambda - u_{0,1}/u_{1,0}), \qquad \lambda \notin \{-1, 0\}, \tag{5.31}
$$

that are of the form[2] $u = f(\tilde{m})$, where $\tilde{m} = m + 2n$.

Solution: With the above ansatz, (5.31) reduces to

$$
f(\tilde{m} + 3) = f(\tilde{m})(\lambda - f(\tilde{m} + 2)/f(\tilde{m} + 1)).
$$

This O∆E has an obvious scaling symmetry, so it can be reduced in order by setting $v_{\tilde{m}} = f(\tilde{m}+1)/f(\tilde{m})$; once $v_{\tilde{m}}$ is known, $f(\tilde{m})$ can be found. The second-order O∆E for $v_{\tilde{m}}$ is

$$
v_{\tilde{m}+2} = (\lambda/v_{\tilde{m}+1} - 1)/v_{\tilde{m}}.
$$

This has no Lie point symmetries, but it is a QRT map, so it has a first integral:

$$
v_{\tilde{m}+1} + v_{\tilde{m}} - 1/v_{\tilde{m}+1} - 1/v_{\tilde{m}} + \lambda/(v_{\tilde{m}+1} v_{\tilde{m}}) = c_1.
$$

At this stage, we can go no further by direct methods.   ▲

Up to this point, we have considered only the wavelike solutions that are invariant in one direction. Another approach that sometimes works is to look for wavelike solutions that are periodic along the wave. A simple ansatz is

$$
u = f(\tilde{m}, \tilde{n}), \qquad (\tilde{m}, \tilde{n}) = (m + kn, n), \tag{5.32}
$$

where f is periodic in its second argument. (One could use any valid lattice map in place of the above shear.)

Example 5.10 Let us seek solutions of the autonomous dpKdV equation,

$$
u_{1,1} = u + \frac{1}{u_{1,0} - u_{0,1}}, \tag{5.33}
$$

that are of the form (5.32), where $k = 1$ and f is of period 2 in $\tilde{n}$. As f has

---

[2] One can determine other wavelike solutions – see Exercise 5.6.

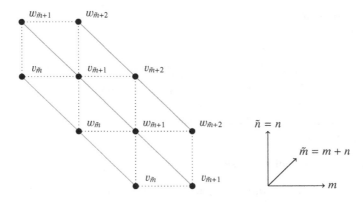

Figure 5.1 Wavelike solution of period 2 along the wave.

period 2, it is helpful to introduce two new unknown functions of $\tilde{m}$ only, as shown in Figure 5.1. Then

$$f(\tilde{m}, \tilde{n}) = \begin{cases} v(\tilde{m}), & \tilde{n} \text{ even}; \\ w(\tilde{m}), & \tilde{n} \text{ odd}. \end{cases}$$

From (5.33), we obtain the following system of OΔEs:

$$v_2 = w - \frac{1}{v_1 - w_1}, \tag{5.34}$$

$$w_2 = v + \frac{1}{v_1 - w_1}. \tag{5.35}$$

This system is easily solved by looking at the sum and difference of (5.34) and (5.35), a calculation that is left as an exercise.   ▲

5.5 How to find Lie symmetries of a given PΔE

For PDEs, every travelling wave solution is invariant under a particular one-parameter Lie group of point symmetries. Other group-invariant solutions can also be found once the Lie point symmetries are known. This suggests that it might be fruitful to seek group-invariant solutions of a given PΔE, that is, solutions on which a characteristic, Q, is zero (see §5.6.1).

Symmetries of a given PΔE are found by a simple generalization of the OΔE method. For example, to find the Lie point symmetries of a PΔE of the form

$$u_{1,1} = \omega(m, n, u, u_{1,0}, u_{0,1}), \tag{5.36}$$

substitute

$$\hat{u} = u + \varepsilon Q(m, n, u) + O(\varepsilon^2)$$

into (5.36) and expand to first order in ε. This yields the LSC,

$$Q(m + 1, n + 1, \omega) = \omega_{,1} Q(m, n, u) + \omega_{,2} Q(m + 1, n, u_{1,0}) + \omega_{,3} Q(m, n + 1, u_{0,1}).$$
$$(5.37)$$

Just as for OΔEs, the LSC is a functional equation, which can be solved by the method of differential elimination.

Example 5.11 For the autonomous dpKdV equation,

$$u_{1,1} = u + \frac{1}{u_{1,0} - u_{0,1}}, \qquad (5.38)$$

the LSC is

$$Q(m + 1, n + 1, \omega) = Q(m, n, u) + \frac{Q(m, n + 1, u_{0,1}) - Q(m + 1, n, u_{1,0})}{(u_{1,0} - u_{0,1})^2}, \qquad (5.39)$$

where ω is the right-hand side of (5.38). Applying the partial differential operator $\partial/\partial u_{1,0} + \partial/\partial u_{0,1}$ to (5.39) yields (after simplification)

$$Q'(m, n + 1, u_{0,1}) - Q'(m + 1, n, u_{1,0}) = 0. \qquad (5.40)$$

Thus $Q''(m + 1, n, u_{1,0}) = 0$ and so, taking (5.40) into account,

$$Q(m, n, u) = A(m + n) u + B(m, n)$$

for some functions A and B. Finally, (5.39) gives the following conditions:

$$A(m + n + 2) = -A(m + n + 1) = A(m + n),$$

$$B(m + 1, n + 1) = B(m, n), \qquad B(m + 1, n) = B(m, n + 1).$$

Therefore the set of all characteristics of Lie point symmetries is spanned by

$$Q_1 = 1, \quad Q_2 = (-1)^{m+n}, \quad Q_3 = (-1)^{m+n} u. \qquad \blacktriangle$$

So far, we have only considered scalar PΔEs with two independent variables. One can also determine the Lie symmetries of a system of L PΔEs, $\mathcal{A} = (\mathcal{A}_1, \ldots, \mathcal{A}_L) = 0$, with M dependent variables and N independent variables[3]. Typically, $L = M$ is needed to determine each u^α; however, there are exceptions, which we shall encounter later.

---

[3] There is an important restriction on $\mathcal{A}$: one must be able (at least, in principle) to switch between writing the equations as $\mathcal{A} = 0$ and writing them in terms of the dependent variables and their shifts. This constraint is satisfied whenever the Jacobian matrix whose components are $\partial \mathcal{A}_l/\partial u_{\mathbf{j}}^\alpha$ has rank L when $\mathcal{A} = 0$, a condition known as the *maximal rank condition* (see Olver, 1993). Every system of PΔEs can be written in a form that satisfies the maximal rank condition; indeed, this is the natural way to write it. For instance, the scalar PΔE $\mathcal{A} \equiv u_{1,0} + u_{0,1} + u = 0$ satisfies the condition, but $\mathcal{A} \equiv (u_{1,0} + u_{0,1} + u)^2 = 0$ does not.

Lie point transformations of a system of PΔEs are of the form

$$\Gamma : (\mathbf{n}, \mathbf{u}) \mapsto (\mathbf{n}, \hat{\mathbf{u}}),$$

where

$$\hat{u}^\alpha = u^\alpha + \varepsilon Q^\alpha(\mathbf{n}, \mathbf{u}) + O(\varepsilon^2), \quad \alpha = 1, \ldots, M. \tag{5.41}$$

The characteristic $\mathbf{Q} = (Q^1, \ldots, Q^M)$ defines a one-parameter local Lie group, $e^{\varepsilon X}$, whose infinitesimal generator is

$$X = Q^\alpha(\mathbf{n}, \mathbf{u}) \, \partial_{u^\alpha}. \tag{5.42}$$

The infinitesimal generator is prolonged to deal with shifted variables in just the same way as for scalar PΔEs:

$$X = (S_{\mathbf{J}} Q^\alpha) \, \partial_{u_{\mathbf{J}}^\alpha} = \sum_{\mathbf{J}} Q^\alpha(\mathbf{n} + \mathbf{J}, \mathbf{u}_{\mathbf{J}}) \partial_{u_{\mathbf{J}}^\alpha}. \tag{5.43}$$

Such transformations are Lie point symmetries if they map the set of solutions of the system of PΔEs to itself. Consequently, the LSC is

$$X \mathcal{A}_l = 0 \quad \text{when} \quad \mathcal{A} = 0, \qquad l = 1, \ldots, L. \tag{5.44}$$

(This condition is derived in the same way as the LSC for scalar PΔEs.)

The generalization to multiple dependent variables is useful for scalar PΔEs that contain a parameter, λ. If we allow transformations that can vary this parameter, λ should be treated as a dependent variable, which, nevertheless, is not determined by the PΔE. Commonly, λ is a mesh parameter that is associated with a discretization of a PDE, in which case $\hat{\lambda}$ usually depends on $\mathbf{n}$ and λ (but not on u). If λ is a uniform parameter, $\hat{\lambda}$ is a function of λ only.

Example 5.12 Find the Lie point symmetries of the PΔE

$$u_{1,1} = u(\lambda - u_{0,1}/u_{1,0}),$$

given that the parameter λ can vary uniformly and $\hat{u}$ is independent of λ.

Solution: In this case, the infinitesimal generator is

$$X = Q^1(m, n, u) \, \partial_u + Q^2(\lambda) \, \partial_\lambda,$$

so the LSC is

$$
\begin{aligned}
Q^1(m + 1, &n + 1, u(\lambda - u_{0,1}/u_{1,0})) \\
&= (\lambda - u_{0,1}/u_{1,0}) \, Q^1(m, n, u) \\
&\quad + \frac{u u_{0,1}}{u_{1,0}^2} \, Q^1(m + 1, n, u_{1,0}) - \frac{u}{u_{1,0}} \, Q^1(m, n + 1, u_{0,1}) + u \, Q^2(\lambda).
\end{aligned}
$$

Differential elimination yields the general solution of the LSC:

$$Q^1(m, n, u) = c_1 u + c_2(-1)^{m+n} u + c_3 nu, \qquad Q^2(\lambda) = c_3 \lambda.$$

There are three linearly independent characteristics, but if we had not allowed λ to vary, we would have found only two. Greater freedom is obtained by relaxing the assumptions about the form of the characteristic. For instance, if Q^1 is allowed to depend on λ, each of the constants c_i is replaced by an arbitrary function of λ. The same approach works for PΔEs with several parameters, each of which may be treated as a dependent variable. ▲

Many PΔEs that arise from applications have generalized Lie symmetries, whose characteristics depend on shifts of the dependent variables. Given **n**, we can regard the infinitesimal generator X of such symmetries as being a vector field on the fibre of continuous variables, $u_{\mathbf{J}}^\alpha$, as follows:

$$X = Q_{\mathbf{J}}^\alpha \, \partial_{u_{\mathbf{J}}^\alpha}. \tag{5.45}$$

The functions $Q_{\mathbf{J}}^\alpha$ are not arbitrary; they are determined by prolongation conditions

$$Q_{\mathbf{J}}^\alpha = S_{\mathbf{J}} Q^\alpha, \qquad \mathbf{J} \in \mathbb{Z}^2, \tag{5.46}$$

where $Q^\alpha = Q_{\mathbf{0}}^\alpha$. There is a useful (and neat) test for determining whether or not a general vector field (5.45) satisfies the prolongation conditions.

Lemma 5.13 *The vector field (5.45) is the prolongation of $Q^\alpha \partial_{u^\alpha}$ if and only if X commutes with each shift operator, S_k.*

Proof Let F be an arbitrary differentiable function of finitely many of the continuous variables $u_{\mathbf{J}}^\alpha$. Then, for any vector field (5.45),

$$S_k(XF) - X(S_k F) = \left(S_k Q_{\mathbf{J}}^\alpha \right) S_k \left(\frac{\partial F}{\partial u_{\mathbf{J}}^\alpha} \right) - Q_{\mathbf{J}}^\alpha \frac{\partial (S_k F)}{\partial u_{\mathbf{J}}^\alpha}$$

$$= \left\{ S_k Q_{\mathbf{J}}^\alpha - Q_{\mathbf{J}+\mathbf{1}_k}^\alpha \right\} \frac{\partial (S_k F)}{\partial u_{\mathbf{J}+\mathbf{1}_k}^\alpha}.$$

As F is arbitrary, X commutes with each S_k if and only if $Q_{\mathbf{J}+\mathbf{1}_k}^\alpha = S_k Q_{\mathbf{J}}^\alpha$ for each k, α and $\mathbf{J}$. By induction, this condition is equivalent to (5.46). □

Corollary 5.14 *If X_1 and X_2 are prolonged vector fields, so is*

$$[X_1, X_2] = X_1 X_2 - X_2 X_1. \tag{5.47}$$

Moreover, any three prolonged vector fields, X_1, X_2 and X_3, satisfy the Jacobi identity:

$$[X_1, [X_2, X_3]] + [X_2, [X_3, X_1]] + [X_3, [X_1, X_2]] = 0. \tag{5.48}$$

Proof   By Lemma 5.13, X_1 and X_2 commute with each S_k, so the commutator $[X_1, X_2]$ also commutes with each S_k and is thus a prolonged vector field. The Jacobi identity holds by definition.   □

The LSC for generalized symmetries is similar to (5.44), but now the given PΔE and its shifted versions, $S_J \mathcal{A} = 0$, must be taken into account. Consequently, the LSC is

$$X\mathcal{A} = 0 \quad \text{when} \quad [\mathcal{A} = 0], \tag{5.49}$$

where the square brackets denote the enclosed expression and all of its shifts. By Lemma 5.13, if X satisfies the LSC for $\mathcal{A} = 0$, it also satisfies the LSC for every PΔE $S_J \mathcal{A} = 0$.

In order to be able to solve (5.49), one must be able to derive a functional equation for $\mathbf{Q}$. This is feasible for scalar PΔEs that can be put into Kovalevskaya form. The same is true for systems of PΔEs. A system of M PΔEs is in Kovalevskaya form if it is of the form

$$\mathcal{A}_\alpha \equiv u^\alpha_{p, \mathbf{K}_\alpha} - \omega^\alpha(\mathbf{n}, \mathcal{I}) = 0, \qquad \alpha = 1, \ldots, M. \tag{5.50}$$

Here $\mathcal{I}$ is the set of initial variables,

$$\mathcal{I} = \left(\mathbf{u}_{\{0, \cdot\}}, \mathbf{u}_{\{1, \cdot\}}, \ldots, \mathbf{u}_{\{p-1, \cdot\}} \right),$$

where $\mathbf{u}_{\{i, \cdot\}}$ denotes a (finite) set of those variables $u^\beta_{\mathbf{J}}$ for which $j^1 = i$. If the domain is $\mathbb{Z}^N$, it is convenient to set $\mathbf{K}_\alpha = \mathbf{0}$, by using shifted dependent variables.

From here on, we restrict attention to (systems of) PΔEs that either are in Kovalevskaya form[4], or can be made so by a change of lattice variables. For such a PΔE, the action (on the set of solutions) of a symmetry that leaves $\mathbf{n}$ invariant is determined by its action on the set $\mathcal{I}$ of initial variables.

Theorem 5.15   *Let $\mathcal{A} = 0$ be a PΔE as described above. If X_1 and X_2 are generalized symmetry generators, so is $[X_1, X_2]$.*

Proof   From the LSC for X_2, there exist functions F_l such that

$$X_2 \mathcal{A}_l = F_l(\mathbf{n}, \mathcal{I}, [\mathcal{A}]), \qquad F_l(\mathbf{n}, \mathcal{I}, [0]) = 0.$$

Consequently,

$$X_1 X_2 \mathcal{A}_l = X_1 \{F_l(\mathbf{n}, \mathcal{I}, [\mathcal{A}]) - F_l(\mathbf{n}, \mathcal{I}, [0])\} = 0 \quad \text{when} \quad [\mathcal{A} = 0];$$

---

[4] This restriction is made for simplicity, but it is more stringent than is necessary. If, for each u^α, one can isolate the highest shift in the n^1-direction, the theorems that hold for Kovalevskaya form are applicable (with appropriate modifications).

the second equality follows from the chain rule and the LSC for X_1 (applied to $\mathcal{A} = 0$ and its shifts). ☐

Two characteristics, $\mathbf{Q}$ and $\tilde{\mathbf{Q}}$, are *equivalent* if $\mathbf{Q} = \tilde{\mathbf{Q}}$ on every solution of the PΔE. If $\mathbf{Q}$ and $\tilde{\mathbf{Q}}$ are equivalent, they give rise to the same set of invariant solutions; moreover, $\mathbf{Q} - \tilde{\mathbf{Q}}$ is a characteristic for trivial generalized Lie symmetries. Thus it is not necessary to obtain more than one characteristic in each equivalence class. Indeed, it is sufficient to seek characteristics that depend only on the independent variables $\mathbf{n}$ and the initial variables $\mathcal{I}$; we call these *basic characteristics*. By the linearity of the LSC, the set $\mathfrak{q}^\infty$ of all basic characteristics is a vector space.

Given a characteristic, $\mathbf{Q} = X\mathbf{u}$, the equivalent basic characteristic is obtained by using the PΔE and its shifts to eliminate all $u_{i,j}^\alpha \notin \mathcal{I}$, an operation that is denoted by $|_\mathcal{I}$. There is a bilinear, antisymmetric bracket on $\mathfrak{q}^\infty$:

$$[\mathbf{Q}_1, \mathbf{Q}_2]_\mathcal{I} = (X_1(\mathbf{Q}_2) - X_2(\mathbf{Q}_1))|_\mathcal{I}, \qquad \mathbf{Q}_1, \mathbf{Q}_2 \in \mathfrak{q}^\infty. \tag{5.51}$$

Theorem 5.16 *The vector space $\mathfrak{q}^\infty$ equipped with the bracket $[\cdot, \cdot]_\mathcal{I}$ is a Lie algebra.*

Proof In the following, each prolonged vector field X_i is determined by a basic characteristic, $\mathbf{Q}_i \in \mathfrak{q}^\infty$, and we define

$$\tilde{\mathbf{Q}}_{i,j} = [X_i, X_j]u, \qquad \mathbf{Q}_{i,j} = \tilde{\mathbf{Q}}_{i,j}|_\mathcal{I} = [\mathbf{Q}_i, \mathbf{Q}_j]_\mathcal{I}.$$

By Theorem 5.15, $[X_i, X_j]$ satisfies the LSC, so $\mathbf{Q}_{i,j} \in \mathfrak{q}^\infty$; in other words, $\mathfrak{q}^\infty$ is closed under the bracket. The LSC for a symmetry generator X_i can be written as $(X_i\mathcal{A})|_\mathcal{I} = 0$, so by the chain rule, $(X_i(\tilde{\mathbf{Q}}_{j,k} - \mathbf{Q}_{j,k}))|_\mathcal{I} = 0$. Hence,

$$\{[X_i, [X_j, X_k]]u\}|_\mathcal{I} = \left\{ X_i\tilde{\mathbf{Q}}_{j,k} - (\mathbf{S_J}\tilde{\mathbf{Q}}_{j,k})\frac{\partial \mathbf{Q}_i}{\partial u_\mathbf{J}} \right\}\bigg|_\mathcal{I} = \left\{ X_i\mathbf{Q}_{j,k} - (\mathbf{S_J}\mathbf{Q}_{j,k})\frac{\partial \mathbf{Q}_i}{\partial u_\mathbf{J}} \right\}\bigg|_\mathcal{I}$$

$$= [\mathbf{Q}_i, \mathbf{Q}_{j,k}]_\mathcal{I} = [\mathbf{Q}_i, [\mathbf{Q}_j, \mathbf{Q}_k]_\mathcal{I}]_\mathcal{I}.$$

This result, together with (5.48), establishes the Jacobi identity for $\mathfrak{q}^\infty$. ☐

Let $\mathfrak{q}^k$ denote the set of all basic characteristics that depend on m, n and the initial variables $u_{i,j}$ that satisfy $|i| + |j| \leq k$. So $\mathfrak{q}^0$ is the set of characteristics of Lie point symmetries, $\mathfrak{q}^1$ is the set of basic characteristics that depend on u at (m, n) and adjacent points, and so on. Clearly, $\mathfrak{q}^k$ is a vector space for each k. With the bracket $[\cdot, \cdot]_\mathcal{I}$, the vector space $\mathfrak{q}^0$ is a Lie algebra, but the other $\mathfrak{q}^k$ need not necessarily be Lie algebras.

We now apply these results to scalar PΔEs of the form

$$u_{1,1} = \omega(m, n, u, u_{1,0}, u_{0,1}), \tag{5.52}$$

whose LSC is

$$S_m S_n Q = \frac{\partial \omega}{\partial u} Q + \frac{\partial \omega}{\partial u_{1,0}} S_m Q + \frac{\partial \omega}{\partial u_{0,1}} S_n Q, \tag{5.53}$$

subject to (5.52) and its shifts. For all such PΔEs, the initial variables are all $u_{i,0}$ and $u_{0,j}$ such that i and j are non-negative. We must pick an ansatz for Q that depends on a finite number of these; one of the simplest is

$$Q = Q(m, n, u, u_{1,0}). \tag{5.54}$$

With this ansatz, (5.53) involves $u, u_{1,0}, u_{2,0}, u_{0,1}, u_{1,1}$ and $u_{2,1}$. To express the LSC is terms of the variables in $\mathcal{I}$, we use

$$u_{2,1} = \omega(m + 1, n, u_{1,0}, u_{2,0}, \omega(m, n, u, u_{1,0}, u_{0,1})) \tag{5.55}$$

and (5.52) to eliminate $u_{2,1}$ and $u_{1,1}$. Having written the LSC as a functional equation, we can solve it by differential elimination.

Example 5.17   The PΔE

$$u_{1,1} = \frac{u_{0,1}(u_{1,0} - 1)}{u} + 1 \tag{5.56}$$

prolongs to the following, whose right-hand side is written in terms of $\mathcal{I}$:

$$u_{2,1} = \frac{\{u_{0,1}(u_{1,0} - 1) + u\}(u_{2,0} - 1)}{u u_{1,0}} + 1.$$

To solve the LSC (5.53) with the ansatz (5.54), evaluate the LSC in terms of $\mathcal{I}$, differentiate the result with respect to $u_{2,0}$ and simplify to obtain

$$\frac{u_{1,1}}{u_{1,1} - 1} Q_{,2}(m + 1, n + 1, u_{1,1}, u_{2,1}) = \frac{u_{1,0}}{u_{1,0} - 1} Q_{,2}(m + 1, n, u_{1,0}, u_{2,0}). \tag{5.57}$$

To eliminate the left-hand side, differentiate with respect to $u_{1,0}$, keeping $u_{1,1}$ and $u_{2,1}$ fixed. This yields the PDE

$$\left\{ \frac{\partial}{\partial u_{1,0}} + \frac{u_{2,0} - 1}{u_{1,0}} \frac{\partial}{\partial u_{2,0}} \right\} \left(\frac{u_{1,0}}{u_{1,0} - 1} Q_{,2}(m + 1, n, u_{1,0}, u_{2,0}) \right) = 0,$$

whose solution, taking (5.57) into account, is

$$Q(m + 1, n, u_{1,0}, u_{2,0}) = (u_{1,0} - 1) A \left(m + 1, \frac{u_{2,0} - 1}{u_{1,0}} \right) + B(m + 1, n, u_{1,0}).$$

Here A is twice differentiable with respect to its continuous argument and B is arbitrary. Having split Q, it is now fairly easy to continue the differential elimination and obtain the general solution of the LSC:

$$Q(m, n, u, u_{1,0}) = f(m)u + \frac{f(m + 1)u(1 - u)}{u_{1,0} - 1}. \tag{5.58}$$

Here f is an arbitrary function, so there is an infinite family of characteristics (5.54). This may seem surprising, given that the PΔE (5.56) has no Lie point symmetries. ▲

Commonly, no new symmetries are found with the ansatz (5.54); it is quite restrictive. Thus one may need to try a higher-order ansatz, such as

$$Q = Q(m, n, u, u_{1,0}, u_{2,0}). \qquad (5.59)$$

Here $u_{3,1}$, $u_{2,1}$ and $u_{1,1}$ are eliminated by using (5.52), (5.55) and

$$u_{3,1} = \omega\big(m + 2, n, u_{2,0}, u_{3,0}, \omega(m + 1, n, u_{1,0}, u_{2,0}, \omega(m, n, u, u_{1,0}, u_{0,1}))\big).$$

Unless ω is very simple, the resulting calculations are horrendous.

Many interesting PΔEs of the form (5.52) can be rearranged as

$$u_{0,1} = \Omega(m, n, u, u_{1,0}, u_{1,1}),$$

and have initial variables $u_{i,j}$ that lie on the cross $ij = 0$. In such cases, it is helpful to replace (5.59) by

$$Q = Q(m, n, u_{-1,0}, u, u_{1,0}), \qquad (5.60)$$

and to use

$$u_{-1,1} = \Omega(m - 1, n, u_{-1,0}, u, u_{0,1})$$

to remove $u_{-1,1}$ from (5.53). Then the function ω will be iterated twice, not three times. Even so, it is not easy to calculate such characteristics. Except in extreme need, it is best not to attempt to solve the LSC by hand; computer algebra is essential, although that too has limits. At the time of writing,

$$Q = Q(m, n, u_{-1,0}, u_{0,-1}, u, u_{1,0}, u_{0,1}) \qquad (5.61)$$

is the most general ansatz[5] that is widely used when the initial variables lie on $ij = 0$. In this case, q^1 is the set of all basic characteristics of the form (5.61).

We illustrate these points by considering two PΔEs that have generalized Lie symmetries. To save forests, the calculations that yield the following results are not shown.

Example 5.18   From Example 5.12, the PΔE

$$u_{1,1} = u(\lambda - u_{0,1}/u_{1,0}), \qquad \lambda > 0, \qquad (5.62)$$

has a two-dimensional Lie algebra of point symmetry generators if λ is fixed. The set of characteristics is spanned by

$$Q_1 = u, \qquad Q_2 = (-1)^{m+n}u.$$

---

[5] In fact, this ansatz simplifies considerably – see (5.75) in Exercise 5.16.

The only generalized symmetry characteristics that have the form (5.54) are point symmetries. The ansatz $Q = Q(m, n, u_{0-1}, u, u_{0,1})$, which is similar to (5.60), produces a new characteristic:

$$Q_3 = \frac{uu_{0,1}}{u_{0,-1}}.$$

By calculating each $[Q_i, Q_j]_{\mathcal{I}}$ one finds that $\mathrm{Span}(Q_1, Q_2, Q_3)$ is an abelian Lie algebra.   ▲

Example 5.19   For the autonomous dpKdV equation,

$$u_{1,1} = u + \frac{1}{u_{1,0} - u_{0,1}}, \tag{5.63}$$

the set $\mathfrak{q}^1$ is spanned by

$$Q_1 = 1, \qquad Q_2 = (-1)^{m+n}, \qquad Q_3 = (-1)^{m+n}u, \qquad Q_4 = \frac{1}{u_{1,0} - u_{-1,0}},$$

$$Q_5 = \frac{1}{u_{0,1} - u_{0,-1}}, \qquad Q_6 = \frac{m}{u_{1,0} - u_{-1,0}} + \frac{n}{u_{0,1} - u_{-1,0}}. \tag{5.64}$$

$\mathrm{Span}(Q_1, Q_2, Q_3, Q_4, Q_5)$ is a Lie algebra, whose nonzero commutators are determined by

$$[Q_1, Q_3]_{\mathcal{I}} = Q_2, \qquad [Q_2, Q_3]_{\mathcal{I}} = Q_1.$$

However, $\mathfrak{q}^1$ is not a Lie algebra, because

$$[Q_4, Q_6]_{\mathcal{I}} = \frac{(u_{2,0} - u_{-2,0})}{(u_{2,0} - u)(u - u_{-2,0})(u_{1,0} - u_{-1,0})^2}$$

is a new characteristic, which belongs to $\mathfrak{q}^2$. Furthermore, there is an infinite hierarchy of distinct basic characteristics (up to equivalence):

$$[Q_4, Q_6]_{\mathcal{I}} \in \mathfrak{q}^2, \quad [Q_4, [Q_4, Q_6]_{\mathcal{I}}]_{\mathcal{I}} \in \mathfrak{q}^3, \quad \Big[Q_4, [Q_4, [Q_4, Q_6]_{\mathcal{I}}]_{\mathcal{I}}\Big]_{\mathcal{I}} \in \mathfrak{q}^4, \dots,$$

each of which is of order two higher than the previous one. Similarly, multiple commutators of Q_5 with Q_6 produce a second infinite hierarchy of basic characteristics. This remarkable behaviour is associated with the fact that the autonomous dpKdV equation is integrable. The characteristic Q_4 is said to be a *mastersymmetry* for Q_6. More generally, a function is a mastersymmetry if (by taking nested commutators, as above) it produces new basic characteristics at each stage, generating an infinite hierarchy of independent symmetries.   ▲

5.6 Some simple uses of symmetries

5.6.1 Invariant solutions

Similarity solutions of PDEs are solutions that are invariant under a group of scaling symmetries. More generally, one can seek solutions that are invariant under any Lie group of symmetries. Just as for ODEs and OΔEs, a solution is invariant under a one-parameter Lie group if the characteristic vanishes on the solution. Exactly the same approach works for PΔEs, as follows.

Example 5.20   In the last example, we found some basic characteristics of

$$u_{1,1} = u + \frac{1}{u_{1,0} - u_{0,1}}. \qquad (5.65)$$

Every point symmetry characteristic that involves u is a multiple of

$$Q = (-1)^{m+n}u + c_1 + c_2(-1)^{m+n}.$$

The invariance condition $Q = 0$ is satisfied if and only if

$$u = c_1(-1)^{m+n+1} - c_2,$$

which implies that $u_{1,0} = u_{0,1}$. Hence, in this example, point symmetries do not yield invariant solutions. However, some generalized symmetries are more fruitful. For instance,

$$Q = 1 - 2(u_{1,0} - u_{-1,0})^{-1}, \qquad c_1 \geq 1,$$

vanishes provided that

$$u = m + f(n) + g(n)(-1)^{m+n},$$

for some functions f and g. Then (5.65) holds if and only if

$$0 = (u_{1,1} - u)(u_{0,1} - u_{1,0}) + 1 = (f_1 - f)^2 - (g_1 - g)^2 - 2(-1)^{m+n}(g_1 - g).$$

Comparing terms with equal dependence on m splits this condition into

$$g_1 - g = 0, \qquad f_1 - f = 0,$$

so we obtain a two-parameter family of invariant solutions,

$$u = m + \tilde{c}_1 + \tilde{c}_2(-1)^{m+n}.$$

This is among the simplest of a rich family of invariant solutions that stem from the characteristics (5.64).   ▲

The search for invariant solutions is limited by the number of independent characteristics that one can calculate. In general, this number is small, though we have met two exceptions. Some PΔEs, such as (5.56), have characteristics that depend on arbitrary functions. If these are (non-point) generalized symmetry characteristics, one can obtain higher-order characteristics merely by taking commutators. The same is true for any PΔE that has a mastersymmetry. The following result reveals another major class of PΔEs with infinitely many linearly independent generalized symmetry characteristics.

Lemma 5.21 *Let* $\mathcal{D}u = 0$ *be a scalar linear homogeneous PΔE on* $\mathbb{Z}^N$. *This PΔE has the characteristic* $Q_1 = u_{\mathbf{J}}$ *if and only if* $\mathbf{S_J}$ *commutes with the linear difference operator* $\mathcal{D}$. *Moreover, if* Q_1 *is a characteristic, so is* $Q_p = (\mathbf{S_J})^k u$, *for each* $k \in \mathbb{Z}$.

Proof   (See Exercise 5.21.)   □

In particular, if all coefficients of the PΔE are constant and the domain is $\mathbb{Z}^2$, Lemma 5.21 implies that $Q = u_{i,j}$ is a characteristic for all $(i, j) \in \mathbb{Z}^2$. Consequently, we may seek solutions of the PΔE that are invariant under any $Q = \alpha^{i,j} u_{i,j}$, where the coefficients $\alpha^{i,j}$ are constant. This can be surprisingly productive.

Example 5.22   The heat equation,

$$u_t = \kappa u_{xx}, \tag{5.66}$$

may be discretized by the following weighted average method:

$$u_{0,1} - u = \mu \left[\theta(u_{1,1} - 2u_{0,1} + u_{-1,1}) + (1 - \theta)(u_{1,0} - 2u + u_{-1,0}) \right]. \tag{5.67}$$

Here $0 \le \theta \le 1$ and $\mu = \kappa \Delta t / (\Delta x)^2$, where $\Delta t, \Delta x$ are (uniform) step lengths in the t- and x-directions, respectively. In particular, $\theta = 0$ gives an explicit numerical scheme, $\theta = 1$ gives a fully implicit scheme and $\theta = 1/2$ gives the Crank–Nicolson scheme. As the PΔE is autonomous, we have a huge choice of characteristics. For simplicity, let us seek solutions that are invariant under

$$Q = u_{2,0} - u.$$

Then

$$u = f(n) + (-1)^m g(n),$$

and so (5.67) reduces to

$$f_1 - f + (-1)^m(g_1 - g) = -4\mu(-1)^m(\theta g_1 + (1 - \theta) g).$$

Equating terms with equal dependence on m yields

$$f_1 = f, \qquad g_1 = \left(\frac{1 - 4\mu(1 - \theta)}{1 + 4\mu\theta}\right)g.$$

Hence

$$f(n) = c_1, \qquad g(n) = c_2\left(\frac{1 - 4\mu(1 - \theta)}{1 + 4\mu\theta}\right)^n.$$

Although the solution for $f(n)$ is not particularly informative, note that $|g(n)|$ grows exponentially with increasing n if

$$\left|\frac{1 - 4\mu(1 - \theta)}{1 + 4\mu\theta}\right| > 1.$$

Thus the invariant solution grows exponentially if $\theta < \frac{1}{2}$ and $\mu > (2 - 4\theta)^{-1}$; otherwise it is bounded. This is precisely the result given by the von Neumann (Fourier series) stability analysis of the weighted average method[6]. ▲

5.6.2 Linearization by a point transformation

As well as yielding invariant solutions, symmetries may be used to detect whether a given nonlinear PΔE is linearizable. Exercise 2.7 showed that if an OΔE is linear or linearizable by a point transformation, its characteristics include all solutions of the linearized (homogeneous) OΔE. The same is true for PΔEs, as illustrated by the following example.

Example 5.23 Every characteristic of Lie point symmetries for

$$u_{1,1} = \frac{uu_{1,0} - 2u_{1,0} + 1}{u - u_{1,0}} \tag{5.68}$$

is of the form

$$Q(m, n, u) = c_1(u - 1) + V(m, n)(u - 1)^2, \tag{5.69}$$

where $v = V(m, n)$ is an arbitrary solution of

$$v_{1,1} = v_{1,0} - v. \tag{5.70}$$

The Lie point symmetries of (5.70) have characteristics (with respect to v),

$$\tilde{Q}(m, n, v) = \tilde{c}_1 v + \tilde{V}(m, n), \tag{5.71}$$

where $v = \tilde{V}(m, n)$ is an arbitrary solution of (5.70). If the point transformation $v = \psi(m, n, u)$ linearizes (5.68) to (5.70), every pair (c_1, V) must correspond

---

[6] Exercise 5.5 examines solutions of (5.67) in more detail.

to a pair $(\tilde{c}_1, \tilde{V})$. The characteristics (5.69) and (5.71) correspond to the same symmetry generator X if they satisfy the change-of-variables formula

$$\tilde{Q}(m, n, v) = X\psi(m, n, u) = Q(m, n, u)\psi'(m, n, u), \qquad (5.72)$$

which amounts to

$$\tilde{c}_1\psi(m, n, u) + \tilde{V}(m, n) = \left(c_1(u - 1) + V(m, n)(u - 1)^2\right)\psi'(m, n, u).$$

This holds for all pairs (c_1, V) if and only if it splits as follows:

$$\tilde{c}_1\psi(m, n, u) = c_1(u - 1)\psi'(m, n, u), \qquad \tilde{V}(m, n) = V(m, n)(u - 1)^2\psi'(m, n, u).$$

It is convenient to choose $\tilde{V}(m, n) = V(m, n)$, which yields

$$\psi(m, n, u) = 1/(1 - u), \qquad \tilde{c}_1 = -c_1.$$

It is easy to check that this point transformation does indeed map (5.68) to (5.70). ▲

This simple method works equally well for nonlinear PΔEs with arbitrarily many independent variables. Whenever the characteristics of Lie point symmetries depend on an arbitrary solution of a linear PΔE (with the same number of independent variables), the linearizing transformation can be found from the change-of-variables formula for characteristics. This is because (locally) every point transformation is a bijection between the set of Lie point symmetries of the original PΔE and the set of Lie point symmetries of the transformed PΔE.

Notes and further reading

Our discussion of wavelike solutions is based on Papageorgiou et al. (1990), which introduced this approach for integrable difference equations. Levi and Yamilov (2011) introduced a simple heuristic test for integrability that is based on the existence of generalized symmetries in q_1. The low-order generalized symmetries of integrable PΔEs typically include mastersymmetries (see, for example, Rasin and Hydon, 2007b; Garifullin and Yamilov, 2012; Xenitidis and Nijhoff, 2012) and recursion operators (see Mikhailov et al., 2011b). A variant of Gardner's method for PDEs provides another way to generate an infinite hierarchy of generalized symmetries (see Rasin and Schiff, 2013).

The wave equation is unusual among scalar PΔEs with two independent variables, because it has two functionally independent first integrals, each of which depends on an arbitrary function of the independent variables (see Example 6.4 for details). This feature is the hallmark of *Darboux integrability*.

Garifullin and Yamilov (2012) includes a classification of Darboux integrable PΔEs, including (5.56) and (5.73).

Where a set of characteristics spans a finite-dimensional Lie algebra, and their generators can be exponentiated, the invariant solutions lie in equivalence classes. For PDEs, these equivalence classes are determined by examining the (adjoint) action of the Lie group on the Lie algebra – see Ovsiannikov (1982) or, for a simplified but more limited version, Hydon (2000b). This technique applies, without modification, to invariant solutions of PΔEs.

Exercises

5.1 Determine the general solution of

$$u_{1,1} + a(m)u_{0,1} + b(n)u_{1,0} + a(m)b(n)u = 0, \qquad a(m)b(n) \neq 0, \ (m,n) \in \mathbb{Z}^2.$$

5.2 (a) Find the solution of

$$u_{1,0} - u_{0,1} + (-1)^{m+n}u = 0, \qquad m \geq 0, n \geq 0,$$

with the general initial condition $u(0,n) = A(n)$.

(b) Evaluate your solution in closed form for $u(0,n) = \cos(n\pi/2)$.

5.3 Use separation of variables to solve the initial-value problem

$$u_{1,0} + u_{-1,0} + u_{0,1} + u_{0,-1} - 4u = 0, \qquad m \geq 0, \ n \in \mathbb{Z},$$

$$u(-1,n) = 0, \qquad u(0,n) = \cos(n\pi/2) + \sin(n\pi/2).$$

[Hint: the initial conditions determine which values of λ are needed.]

5.4 Consider the general linear homogeneous PΔE of the form

$$a^{1,0}(m,n)u_{1,0} + a^{0,1}(m,n)u_{0,1} + a^{0,0}(m,n)u = 0, \qquad (m,n) \in D,$$

where the functions $a^{i,j}(m,n)$ are nonzero for all $(m,n) \in D$. Under what conditions on $a^{i,j}(m,n)$ does the PΔE have nonzero separable solutions? Extend your result to PΔEs of the form

$$a^{1,1}(m,n)u_{1,1} + a^{1,0}(m,n)u_{1,0} + a^{0,1}(m,n)u_{0,1} + a^{0,0}(m,n)u = 0.$$

5.5 (a) Find all separable solutions of the weighted average approximation (5.67) that are bounded as $|m| \to \infty$.

(b) Under what conditions do such solutions stay bounded as $n \to \infty$?

(c) What do these results imply about the stability of the weighted average approximation to the heat equation (5.66)?

5.6 (a) Write down all solutions of (5.30) that are of the form $u = f(m - n)$.

    (b) Discover as many wavelike solutions of (5.31) as you can.

5.7 (a) Which wavelike solutions of (5.67) are of the form $u = f(m - n)$?

    (b) When do these solutions grow without limit as $m - n \to \infty$?

    (c) Investigate other wavelike solutions of (5.67).

5.8 Determine the general solution of the system (5.34) and (5.35).

5.9 Find all characteristics for Lie point symmetries of the PΔE

$$u_{1,1} = u_{0,1} + \frac{1}{u_{1,0} - u}. \qquad (5.73)$$

[Although this PΔE looks similar to the autonomous dpKdV equation, it has very different properties, as it is Darboux integrable.]

5.10 For the wave equation (5.7), determine all generalized symmetry characteristics that are of the form $Q(m, n, u, u_{1,0})$. Why does the general solution include an arbitrary function that involves the dependent variables?

5.11 Solve the LSC in Example 5.12.

5.12 Show that (5.58) is the general solution of the LSC for (5.56) that satisfies the ansatz (5.54).

5.13 Show that the PΔE

$$u_{1,1} + 1/u_{1,0} + 1/u_{0,1} + u = 0$$

has a three-dimensional vector space of characteristics that are of the form $Q(m, n, u_{-1,0}, u, u_{1,0})$. Is this vector space (with the bracket $[\cdot, \cdot]_\mathcal{I}$) a Lie algebra? Can you obtain further symmetries of the PΔE?

5.14 (a) Show that the ansatz (5.60) yields the following characteristics for the PΔE (5.62):

$$Q_4 = \frac{uu_{1,0}}{u_{1,0} - u_{-1,0}}, \qquad Q_5 = (m + n)u - \frac{2muu_{1,0}}{u_{1,0} - u_{-1,0}}.$$

    (b) Prove that, for this PΔE, $\mathfrak{q}^1 = \mathrm{Span}(Q_1, \ldots, Q_5)$.

    (c) Calculate $[Q_4, Q_5]_\mathcal{I}$ and determine whether Q_4 or Q_5 is a master-symmetry.

5.15 The *cross-ratio equation* is

$$\alpha(uu_{1,0} + u_{0,1}u_{1,1}) - \beta(uu_{0,1} + u_{1,0}u_{1,1}) = 0, \qquad (5.74)$$

where α and β are parameters. Find all symmetry generators of the form

$$X = Q^u(m, n, u_{-1,0}, u_{0,-1}, u, u_{1,0}, u_{0,1}, \alpha, \beta)\partial_u + Q^\alpha(m, \alpha)\partial_\alpha + Q^\beta(n, \beta)\partial_\beta.$$

Are any of the corresponding characteristics mastersymmetries?

5.16 Suppose that each $u_{i,j}$ in (5.52) can be isolated, so that generic initial conditions on the cross $ij = 0$ determine the solution on $\mathbb{Z}^2$ uniquely. Prove that every characteristic (5.61) in q_1 is of the following split form:

$$Q = A(m, n, u_{-1,0}, u, u_{1,0}) + B(m, n, u_{0,-1}, u, u_{0,1}). \qquad (5.75)$$

[If you had not noticed this already, look again at the worked examples on generalized Lie symmetries.] This result is the reason why (5.61) is a manageable ansatz!

5.17 Find all solutions of (5.62) that are invariant under the symmetries whose characteristic is $4Q_3 - \lambda^2 Q_1$.

5.18 The heat equation (5.66) may be approximated by the explicit scheme

$$u_{0,1} - u = \mu(u_{1,0} - 2u + u_{-1,0}), \qquad m \in \mathbb{Z}, \ n \geq 0. \qquad (5.76)$$

Identify the appropriate initial line and (assuming that $0 < \mu < 1/2$), determine all basic characteristics in q^1 that are linear in m and n. Find all solutions of (5.76) that are invariant under one of these characteristics.

5.19 Show that

$$Q = (u_{1,0} + 2u + u_{-1,0} + 2\alpha)(u_{1,0} - u_{-1,0})^{-1}$$

is a basic characteristic for the PΔE

$$(u_{1,1} - u)(u_{0,1} - u_{1,0}) + (\beta - \alpha)(u_{1,1} + u_{0,1} + u_{1,0} + u) + \beta^2 - \alpha^2 = 0,$$

where α, β are parameters. Use this characteristic to obtain a family of invariant solutions of the PΔE.

5.20 Show that the autonomous dpKdV equation (5.63) has a family of solutions,

$$u = c_1 2^{(1-n)/2} \cos\left((m + n)\pi/2\right) + c_1^{-1} 2^{n/2} \sin\left((m + n)\pi/2\right),$$

and that these solutions are invariant under $[Q_4, Q_6]_{\mathcal{I}}$. Find all other families of solutions that are invariant under this characteristic.

5.21 Examine the effect of a given stencil translation on the LSC for a generalized symmetry. Hence prove Lemma 5.21.

5.22 (a) Find the set of all characteristics of Lie point symmetries for

$$u_{1,1} = \frac{uu_{1,0}u_{0,1} + u_{1,0} + u_{0,1} - u}{uu_{1,0} + uu_{0,1} - u_{1,0}u_{0,1} + 1}. \tag{5.77}$$

   (b) Use your result to construct a point transformation that linearizes (5.77).

   (c) Determine the general solution of (5.77).

6

Conservation laws

Though analogy is often misleading,
it is the least misleading thing we have.

(Samuel Butler, Notebooks)

A conservation law is an expression of the behaviour of all solutions of a given PΔE; it is a direct consequence of the PΔE. This enables one to derive all conservation laws that satisfy a given (sufficiently simple) ansatz, even if no solutions are known. If the PΔE is the Euler–Lagrange equation for a variational problem, conservation laws are linked with variational symmetries, because Noether's Theorem for PDEs extends to PΔEs. Similarly, there is an analogue of Noether's (lesser-known) Second Theorem, which deals with gauge symmetries. This chapter describes each of these approaches, together with a useful result that bridges the gap between Noether's two theorems.

6.1 Introduction to conservation laws

Let $\mathcal{A} \equiv (\mathcal{A}_1, \ldots, \mathcal{A}_L) = 0$ be a PΔE that has N independent variables. A *conservation law* for this PΔE is an expression of the form

$$\mathcal{C}(\mathbf{n}, [\mathbf{u}]) \equiv (S_1 - I)F^1(\mathbf{n}, [\mathbf{u}]) + \cdots + (S_N - I)F^N(\mathbf{n}, [\mathbf{u}]) \qquad (6.1)$$

that vanishes on solutions of the PΔE. In other words,

$$\mathcal{C}(\mathbf{n}, [\mathbf{u}]) = 0 \quad \text{when} \quad [\mathcal{A} = 0]. \qquad (6.2)$$

Usually, we will drop the qualifier 'when $[\mathcal{A} = 0]$'. For instance, conservation laws for a scalar PΔE with two independent variables will be written as

$$(S_m - I)F(m, n, [u]) + (S_n - I)G(m, n, [u]) = 0. \qquad (6.3)$$

Conservation laws of a given PDE are defined similarly, but with total derivatives replacing the difference operators $(S_i - I)$; in other words, they are divergence expressions that vanish on solutions of the PDE. By analogy, any expression of the form (6.1) is called a (difference) divergence.

165

Example 6.1 The autonomous dpKdV equation,

$$u_{1,1} = u + 1/(u_{1,0} - u_{0,1}), \tag{6.4}$$

has (infinitely) many conservation laws, including

$$(S_m - I)\{(-1)^{m+n}(1 - 2uu_{0,1})\} + (S_n - I)\{2(-1)^{m+n}uu_{1,0}\} = 0. \tag{6.5}$$

To see that (6.5) is a conservation law, expand its left-hand side to get

$$\mathcal{C}(m, n, [u]) = 2(-1)^{m+n}\{(u_{1,1} - u)(u_{1,0} - u_{0,1}) - 1\},$$

which is zero on all solutions of (6.4). ▲

The condition (6.2) is linear homogeneous in $\mathbf{F} = (F^1, \dots, F^N)$, so the set of all conservation laws of a given PΔE is a vector space. A common convention is to describe $\mathbf{F}$ itself as being the conservation law; although this is a slight abuse of notation, we will adopt it whenever the meaning is obvious.

Just as for PDEs, some conservation laws are trivial – they give no real information about the PΔE. For instance, if all components F^i vanish on solutions of the PΔE, (6.2) holds automatically. This is the first type of triviality, which is suppressed by requiring that the stencil of each F^i contains no translate of the stencil, S, for the PΔE. In particular, if the PΔE can be put into Kovalevskaya form, one usually requires that $\mathbf{F}$ depends only on $\mathbf{n}$ and the initial variables $\mathcal{I}$.

The second type of triviality occurs when (6.2) is satisfied identically, without reference to the PΔE. For instance,

$$F = (S_n - I) f(m, n, [u]), \qquad G = -(S_m - I) f(m, n, [u]),$$

is a trivial conservation law for any function f, because (6.3) is satisfied identically. Similarly, any set of functions f^{ij}, $i < j$, gives rise to a trivial conservation law:

$$F^i = \sum_{j=1}^{N} (S_j - I) f^{ij}(\mathbf{n}, [\mathbf{u}]), \qquad \text{where} \quad f^{ij} = -f^{ji}, \quad i \geq j. \tag{6.6}$$

For simplicity, we will restrict attention throughout this chapter to difference equations on regular domains (usually $\mathbb{Z}^N$). As such domains have no 'holes' (missing interior points), the second type of triviality occurs if and only if[1] (6.6) is satisfied for some functions f^{ij}.

With the above restriction, a conservation law is trivial if and only if it is a linear superposition of the two types of trivial conservation laws discussed

---

[1] This result is established with the aid of difference cohomology, which is also able to deal with domains that are not product lattices – see Mansfield and Hydon (2008).

above. Two conservation laws are *equivalent* if they differ by a trivial conservation law. We need to find a basis for the set of equivalence classes. (For $N = 1$, any nontrivial conservation law is a first integral; the problem of finding these was discussed in Chapter 2.) One difficulty is that we do not know in advance how many linearly independent equivalence classes exist. A pragmatic approach is to ignore this difficulty and merely to try to find as many nontrivial conservation laws as possible, by whatever means one can.

6.2 Conservation laws for linear PΔEs

In this section, we consider conservation laws of a given linear homogeneous scalar PΔE on $\mathbb{Z}^N$. The ideas are easily extended to inhomogeneous linear PΔEs and to domains other than $\mathbb{Z}^N$.

The ℓ^2 *inner product* of two real-valued functions u, v on $\mathbb{Z}^N$ is

$$\langle u, v \rangle = \sum_{\mathbf{n} \in \mathbb{Z}^N} u(\mathbf{n}) v(\mathbf{n}). \tag{6.7}$$

Clearly, this is defined only for functions for which the right-hand side of (6.7) converges. Let $\mathcal{D}$ be a linear difference operator. Then the *adjoint operator* $\mathcal{D}^\dagger$ is determined by

$$\langle u, \mathcal{D}^\dagger v \rangle = \langle \mathcal{D}u, v \rangle \tag{6.8}$$

for all u, v such that the inner products are defined. From here on, we shall work formally, ignoring any issues that are associated with convergence.

Lemma 6.2 *Given a linear homogeneous scalar PΔE,*

$$\mathcal{D}u = 0, \tag{6.9}$$

suppose that the adjoint PΔE, $\mathcal{D}^\dagger v = 0$, *has a nonzero solution,* $v = V(\mathbf{n})$. *Then*

$$V(\mathbf{n}) \mathcal{D}u = 0 \tag{6.10}$$

can be rearranged as a conservation law for (6.9).

Proof   To keep the notation concise, we consider only the case $N = 2$; the generalization to arbitrary N is obvious. Let

$$\mathcal{D}u = \sum_{(i,j) \in \mathcal{S}} a^{i,j}(m, n) u_{i,j}.$$

Translating the stencil if necessary, we will assume (without loss of generality) that $i \geq 0$ and $j \geq 0$ in the above sum. For each $(i, j) \in \mathcal{S}$,

$$\sum_{m,n\in\mathbb{Z}} a^{i,j}(m, n) u_{i,j} v = \sum_{m,n\in\mathbb{Z}} a^{i,j}(m - i, n - j) u v_{-i,-j},$$

which (formally) is a consequence of the identity

$$a^{i,j}(m, n) u_{i,j} v \; - \left(S_m - I\right)\left\{\sum_{k=0}^{i-1} a^{i,j}(m - i + k, n) u_{k,j} v_{k-i,0}\right\} \tag{6.11}$$

$$= a^{i,j}(m - i, n - j) u v_{-i,-j} + \left(S_n - I\right)\left\{\sum_{l=0}^{j-1} a^{i,j}(m - i, n - j + l) u_{0,l} v_{-i,l-j}\right\}.$$

Thus the adjoint PΔE is

$$\mathcal{D}^{\dagger} v \equiv \sum_{(i,j)\in\mathcal{S}} a^{i,j}(m - i, n - j) v_{-i,-j} = 0. \tag{6.12}$$

By summing (6.11) over all $(i, j) \in \mathcal{S}$ and rearranging, we obtain an expression of the form

$$v\left(\mathcal{D}u\right) - u\left(\mathcal{D}^{\dagger}v\right) = (S_m - I)F + (S_n - I)G. \tag{6.13}$$

Evaluating (6.13) on a nonzero solution of the adjoint PΔE gives the required result.   □

Lemma 6.2 is very easy to use in practice: first find F and G from (6.13), then replace v by a solution of the adjoint PΔE.

Example 6.3   Consider the PΔE

$$\mathcal{D}u \equiv u_{1,0} - u_{0,1} + u = 0, \tag{6.14}$$

whose adjoint is

$$\mathcal{D}^{\dagger} v \equiv v_{-1,0} - v_{0,-1} + v = 0.$$

In this case, (6.13) is

$$v\left(\mathcal{D}u\right) - u\left(\mathcal{D}^{\dagger}v\right) = (S_m - I)\left(v_{-1,0}u\right) + (S_n - I)\left(-v_{0,-1}u\right).$$

For the simple solution $v = 2^{-n}$, the resulting conservation law for (6.14) is

$$(S_m - I)\left(2^{-n}u\right) + (S_n - I)\left(-2^{1-n}u\right) = 0.$$

Similarly, the general separable solution $v = \mu^{m+n}(1 + \mu)^{-n}$ yields

$$(S_m - I)\left(\mu^{m+n-1}(1 + \mu)^{-n}u\right) + (S_n - I)\left(-\mu^{m+n-1}(1 + \mu)^{1-n}u\right) = 0. \quad \blacktriangle$$

Example 6.4 The adjoint of the wave equation,

$$\mathcal{D}u \equiv u_{1,1} - u_{1,0} - u_{0,1} + u = 0, \tag{6.15}$$

is the same equation, but with the stencil translated one step back in m and n:

$$\mathcal{D}^\dagger v \equiv v_{-1,-1} - v_{-1,0} - v_{0,-1} + v = 0. \tag{6.16}$$

The conservation law (6.10) corresponding to any particular solution of (6.16) is

$$(S_m - I)\{(v_{-1,-1} - v_{-1,0})u\} + (S_n - I)\{v_{0,-1}(u_{1,0} - u)\} = 0.$$

As the general solution of (6.16), $v = f(m) + g(n)$, depends on arbitrary functions, the wave equation has an infinite family of conservation laws:

$$(S_m - I)\{(g(n-1) - g(n))u\} + (S_n - I)\{(f(m) + g(n-1))(u_{1,0} - u)\} = 0.$$

An equivalent (but more symmetric) form is

$$(S_m - I)\{g(n)(u_{0,1} - u)\} + (S_n - I)\{f(m)(u_{1,0} - u)\} = 0. \tag{6.17}$$

A *first integral* of a given PΔE is a function $\phi(\mathbf{n}, [\mathbf{u}])$ such that $(S_\mathbf{J} - I)\phi = 0$ is a nontrivial conservation law, for at least one $\mathbf{J}$ that is visible from $\mathbf{0}$. The wave equation (6.15) is unusual, because (6.17) yields two families of first integrals, each of which depends on an arbitrary function of the independent variables[2], namely $\phi = f(m)(u_{1,0} - u)$ and $\tilde{\phi} = g(n)(u_{0,1} - u)$. ▲

6.3 The direct method

The most direct way of discovering conservation laws is to choose an ansatz for $\mathbf{F}$, then find the circumstances under which this postulated conservation law vanishes on solutions of the PΔE. For PΔEs with two independent variables, conservation laws are of the form

$$(S_m - I)F + (S_n - I)G = 0. \tag{6.18}$$

In this section, we will restrict attention to scalar PΔEs of the form

$$u_{1,1} = \omega(m, n, u, u_{1,0}, u_{0,1}). \tag{6.19}$$

The PΔE (6.19) and its shifts will be used to write (6.18) in terms of (m, n) and the set $\mathcal{I}$ of initial variables, creating a functional equation that determines F and G. This idea extends immediately to all PΔEs that can be written in Kovalevskaya form.

---

[2] This is a feature of Darboux integrable PΔEs; see Notes and further reading in Chapter 5.

The simplest nontrivial conservation laws are those that involve only m, n and the values $u_{i,j}$ that occur in (6.19). Up to equivalence, the most general such conservation law has components of the form

$$F = F(m, n, u, u_{0,1}), \quad G = G(m, n, u, u_{1,0}). \qquad (6.20)$$

In order to keep the notation concise, F, G and ω will stand for these functions with the arguments given above.

The conservation laws (6.20) are calculated by evaluating (6.18) on $\mathcal{I}$; when F and G are of the form (6.20), only the initial variables u, $u_{1,0}$ and $u_{0,1}$ are needed. This gives the *determining equation*,

$$(S_m F)|_{\mathcal{I}} - F + (S_n G)|_{\mathcal{I}} - G \equiv 0, \qquad (6.21)$$

which is shorthand for

$$F(m+1, n, u_{1,0}, \omega) - F(m, n, u, u_{0,1}) + G(m, n+1, u_{0,1}, \omega) - G(m, n, u, u_{1,0}) \equiv 0.$$

Although the functional equation (6.21) depends on the three initial variables, each function has only two continuous arguments. Therefore, assuming that F and G are locally smooth in their continuous arguments, (6.21) may be solved by differential elimination.

Our assumptions about F and G preclude the first type of triviality. However, the second type occurs if F contains a term of the form $(S_n - I)f(m, n, u)$. Indeed, any term in F that is independent of u can be written as

$$f(m, n+1, u_{0,1}) = f(m, n, u) + (S_n - I)f(m, n, u).$$

Thus, to classify the conservation laws up to equivalence, we will remove from F all terms that are independent of u. Terms in G that are of the form $f(m)$ will also be removed, as they do not appear in the conservation law (6.18).

Example 6.5   Up to equivalence, find all conservation laws (6.20) for

$$u_{1,1} = u_{0,1} + \frac{1}{u_{0,1}} - \frac{1}{u_{1,0}}, \qquad m \geq 0, \ n \geq 0. \qquad (6.22)$$

Solution: For this PΔE, ω is independent of u. By differentiating (6.21) with respect to u, we can eliminate two terms at once, obtaining

$$F_{,1} + G_{,1} = 0. \qquad (6.23)$$

Next, eliminate $G_{,1}$ by differentiating with respect to $u_{0,1}$, which gives $F_{,12} = 0$. Solving this PΔE and removing terms that are independent of u, we obtain

$$F(m, n, u, u_{0,1}) = A(m, n, u),$$

where A is arbitrary (but locally smooth in u). Then (6.23) yields

$$G(m, n, u, u_{1,0}) = -A(m, n, u) + B(m, n, u_{1,0}),$$

so the determining equation (6.21) amounts to

$$A(m + 1, n, u_{1,0}) - A(m, n + 1, u_{0,1}) + B(m, n + 1, \omega) - B(m, n, u_{1,0}) = 0. \quad (6.24)$$

Two terms can be eliminated by differentiating with respect to $u_{0,1}$:

$$\left(1 - 1/u_{0,1}^2\right) B'(m, n + 1, \omega) - A'(m, n + 1, u_{0,1}) = 0.$$

Thus, discounting contributions to A that are independent of u,

$$A(m, n, u) = a(m, n)(u + 1/u), \qquad B(m, n, u_{1,0}) = a(m, n)u_{1,0} + b(m, n).$$

These results reduce (6.24) to

$$(a_{1,0} - a)u_{1,0} + (a_{1,0} - a_{0,1})/u_{1,0} + b_{0,1} - b = 0.$$

Comparing terms with the same dependence on $u_{1,0}$, we obtain the solution

$$a(m, n) = c_1, \qquad b(m, n) = f(m);$$

without loss of generality, we can set $f = 0$. Thus, up to equivalence, every conservation law (6.20) for the PΔE (6.22) is a multiple of

$$(F_1, G_1) = (u + 1/u, \quad u_{1,0} - u - 1/u). \qquad \blacktriangle$$

The above example is exceptionally easy; the entire calculation takes just a few lines. Typically, a single differentiation does not eliminate more than one term. To keep expression swell to a minimum, it is usual to begin by eliminating the terms containing ω from (6.21). This is achieved by applying each of the following differential operators to (6.21):

$$L_1 = \frac{\partial}{\partial u_{1,0}} - \frac{\omega_{,2}}{\omega_{,1}} \frac{\partial}{\partial u}, \qquad L_2 = \frac{\partial}{\partial u_{0,1}} - \frac{\omega_{,3}}{\omega_{,1}} \frac{\partial}{\partial u}.$$

These operators commute, so

$$L_1 L_2 \{F(m + 1, n, u_{1,0}, \omega) + G(m, n + 1, u_{0,1}, \omega)\} = 0.$$

(The elimination of both terms containing ω is possible in two steps only because L_1 and L_2 commute. Non-commuting differential operators generate extra terms, increasing the rate of expression swell.) This elimination produces the functional-differential equation

$$-L_1 L_2(F + G) = 0,$$

which can be rewritten as

$$L_2\left(\frac{\omega_{,2}}{\omega_{,1}}F_{,1}\right) + L_1\left(\frac{\omega_{,3}}{\omega_{,1}}G_{,1}\right) = 0. \tag{6.25}$$

Depending on ω, either two or three terms in (6.25) involve derivatives of G. They are eliminated in the usual way, by successively isolating each term and differentiating with respect to $u_{0,1}$. Hence (6.25) yields a PDE for F that is at most fifth-order. As we know which differential operators have been applied to obtain this PΔE, we can reverse the process to obtain F (and hence, G), up to unknown functions of m and n. These functions are found by going up the hierarchy of functional-differential equations until (6.21) is satisfied.

As this process generates very lengthy expressions, it is essential to use computer algebra for all but the simplest problems. Under some circumstances, the efficiency of the computations can be increased. In particular, if it is possible to rewrite the PDE (6.19) in the form

$$u_{1,0} = \Omega(m, n, u, u_{0,1}, u_{1,1}), \tag{6.26}$$

the conservation law (6.18) amounts to

$$F(m+1, n, \Omega, u_{1,1}) - F(m, n, u, u_{0,1}) + G(m, n+1, u_{0,1}, u_{1,1}) - G(m, n, u, \Omega) = 0. \tag{6.27}$$

By applying the commuting differential operators

$$L_3 = \frac{\partial}{\partial u} - \frac{\Omega_{,1}}{\Omega_{,2}}\frac{\partial}{\partial u_{0,1}}, \quad L_4 = \frac{\partial}{\partial u_{1,1}} - \frac{\Omega_{,3}}{\Omega_{,2}}\frac{\partial}{\partial u_{0,1}},$$

one can reduce (6.27) to

$$L_3 L_4\left(-F + G(m, n+1, u_{0,1}, u_{1,1})\right) = 0. \tag{6.28}$$

Again, up to three differentiations are needed to derive a PDE for F from (6.28). Perhaps surprisingly, the information that is obtained from this PDE may differ from what can be found from (6.25).

Example 6.6   The autonomous dpKdV equation,

$$u_{1,1} = u + \frac{1}{u_{1,0} - u_{0,1}}, \tag{6.29}$$

can be written in either of the forms (6.19) and (6.26), with

$$\omega = u + \frac{1}{u_{1,0} - u_{0,1}}, \quad \Omega = u_{0,1} + \frac{1}{u_{1,1} - u}.$$

Skipping details of the calculations, the first approach (using ω) yields the PDE

$$(u_{1,0} - u_{0,1})^2 F_{,12222} - 4(u_{1,0} - u_{0,1})F_{,1222} - F_{,11222} = 0.$$

Bearing in mind that F is independent of $u_{1,0}$, this simplifies to

$$F_{,1222} = 0. \tag{6.30}$$

The second approach (in which $u_{1,0}$ is replaced by Ω) gives

$$(u_{1,1} - u)^2 F_{,11112} - 4(u_{1,1} - u)F_{,1112} - F_{,11122} = 0,$$

which simplifies to[3]

$$F_{,1112} = 0. \tag{6.31}$$

By using both (6.30) and (6.31), one obtains F with far less computational effort than is needed if (6.30) alone is used. A routine calculation (which is left to the reader) completes the solution of the determining equation. Up to equivalence, the conservation laws are spanned by

$$(F_1, G_1) = \left((-1)^{m+n} u u_{0,1}, \; (-1)^{m+n+1}(u u_{1,0} + 1/2)\right);$$

$$(F_2, G_2) = \left(u u_{0,1}^2 - u^2 u_{0,1}, \; u^2 u_{1,0} - u u_{1,0}^2 + u - u_{1,0}\right);$$

$$(F_3, G_3) = \left((-1)^{m+n} u u_{0,1}(u + u_{0,1}), \; (-1)^{m+n+1}\left(u^2 u_{1,0} + u u_{1,0}^2 + u + u_{1,0}\right)\right);$$

$$(F_4, G_4) = \left((-1)^{m+n} u u_{0,1}(u u_{0,1} - 2), \; (-1)^{m+n+1}\left(u^2 u_{1,0}^2 - 1/2\right)\right). \qquad \blacktriangle$$

So far, we have looked for conservation laws whose components depend on three initial values; these are known as *three-point conservation laws*. Higher-order conservation laws may be found by allowing each F^i to depend on more initial variables. For instance, in the scheme shown in Figure 6.1(a), the components are of the form

$$F = F(m, n, u_{-1,0}, u_{-1,1}, u_{0,-1}, u), \qquad G = G(m, n, u_{-1,0}, u_{0,-1}, u, u_{1,-1}).$$

These are examples of *five-point conservation laws*: shifted versions of the PΔE (6.19) are used to eliminate $u_{0,1}$ and $u_{1,0}$, leaving a determining equation that depends on five initial values of u that lie on the staircase.

Given any three-point conservation law with components (F_1, G_1), there is a family of five-point conservation laws on the staircase whose components are

$$F = c_1 S_m^{-1} F_1 + c_2 S_n^{-1} F_1, \qquad G = c_1 S_m^{-1} G_1 + c_2 S_n^{-1} G_1.$$

(The original three-point conservation law is used on each 'stair'.) Each of these conservation laws is equivalent to a multiple of the original one. Consequently, it is reasonable to restrict the term 'five-point' to equivalence classes of conservation laws that are not equivalent to three-point conservation laws.

---

[3] There is a neat way to obtain (6.31) directly from (6.30): simply observe that the autonomous dpKdV equation has a discrete symmetry which interchanges the values $u_{i,0}$ and $u_{i,1}$ for each i.

Figure 6.1 Five-point conservation laws on: (a) the staircase; (b) the cross. The values of u at solid points appear in the determining equation; values at ringed points are eliminated. Horizontal (resp. vertical) arrows indicate the shift in m (resp. n).

If (6.19) can also be written in the forms (6.26) and

$$u_{0,1} = \tilde{\Omega}(m, n, u, u_{1,0}, u_{1,1}),$$

we may use five initial points on the cross $ij = 0$. In this case,

$$F = F(m, n, u_{-1,0}, u_{0,-1}, u, u_{0,1}), \quad G = G(m, n, u_{0,-1}, u_{-1,0}, u, u_{1,0}). \quad (6.32)$$

The two approaches yield the same five-point conservation laws (up to equivalence). However, it is computationally more efficient to use initial values on the cross, because Ω and $\tilde{\Omega}$ each occur in just one term of the determining equation, as indicated in Figure 6.1(b).

Example 6.7 Up to equivalence modulo three-point and trivial conservation laws, the ansatz (6.32) yields the following basis for the five-point conservation laws of the autonomous dpKdV equation (6.29):

$$(F_5, G_5) = \Big(\ln(u_{0,1} - u_{-1,0}), \ -\ln(u_{1,0} - u_{-1,0}) \Big); \quad (6.33)$$

$$(F_6, G_6) = \Big(\ln(u_{0,1} - u_{0,-1}), \ -\ln(u_{1,0} - u_{0,-1}) \Big);$$

$$(F_7, G_7) = m(F_5, G_5) + n(F_6, G_6). \qquad \blacktriangle$$

6.4 When are two conservation laws equivalent?

In principle, higher-order conservation laws may be calculated by the direct method. For all but the simplest PΔEs, expression swell makes this impracticable. However, there are other methods of finding conservation laws that are not limited in this way. A question arises immediately. How can we tell whether an apparently new conservation law is equivalent to a conservation law that is already known?

Example 6.8 The conservation laws of the *discrete Lotka–Volterra (dLV)* equation,

$$u_{1,1} = \frac{u\,(u_{0,1} - 1)}{u_{1,0} - 1},$$ (6.34)

include

$$(F_1, G_1) = (u(u_{0,1} - 1),\ -u_{1,0}),$$ (6.35)

$$(F_2, G_2) = \left(u_{0,1}(u_{0,2} + u - 2),\ \frac{u\,(u_{1,0} + u_{0,1} - 2)}{1 - u_{1,0}}\right).$$ (6.36)

It appears that (F_2, G_2) is of higher order than (F_1, G_1). In fact, it is equivalent to twice (F_1, G_1), because

$$F_2 = 2F_1 + (S_n - I)(u(u_{0,1} - 2)),$$
$$G_2 = 2G_1 - (S_m - I)(u(u_{0,1} - 2)) + u_{1,0}(u_{1,1} - \omega),$$

where ω is the right-hand side of (6.34). Even though (6.35) and (6.36) each lie on the initial lines $ij = 0$, both types of triviality are needed to prove this equivalence if one works in terms of the components. It is instructive to write out the conservation laws in full, which immediately eliminates the second type of triviality, and to use $\mathcal{A} = u_{1,1} - \omega$:

$$\mathcal{C}_1 = (u_{1,0} - 1)\mathcal{A},$$
$$\mathcal{C}_2 = (\mathcal{A} + \omega)\,S_n\mathcal{A} + (u_{1,0} - 2)\mathcal{A}$$
$$= 2(u_{1,0} - 1)\mathcal{A} + (S_n - I)(u_{1,0}\mathcal{A}).$$

The term $(S_n - I)(u_{1,0}\mathcal{A})$ gives the first type of triviality; it corresponds to the appearance of $u_{1,0}\mathcal{A}$ in G_2. Thus it is obvious that $\mathcal{C}_2$ is equivalent to $2\mathcal{C}_1$. ▲

For PDEs, every conservation law is associated with a characteristic function that can be used to test for equivalence. It turns out that there is a parallel (but more complicated) structure for PΔEs. Any conservation law of the form

$$\mathcal{C} = \mathcal{A}_l \mathcal{Q}^l,$$ (6.37)

where $\mathcal{Q} = (\mathcal{Q}^1, \dots, \mathcal{Q}^L)$ has a finite limit (for generic initial data) as $\mathcal{A} \to 0$, is said to be in *characteristic form*. The L-tuple $\mathcal{Q}$ is called the *characteristic* of the conservation law. For instance, $\mathcal{C}_1$ in Example 6.8 is in characteristic form; its characteristic is $\mathcal{Q} = u_{1,0} - 1$. However, the equivalent conservation law $\mathcal{C}_2$ is not in characteristic form. Even worse, the conservation laws in Example 6.7 involve logarithms of rational functions of $\mathcal{A}$ and its shifts (see Exercise 6.6). Is it possible to write each of these (or an equivalent conservation law) in characteristic form?

For any PΔE that can be written in Kovalevskaya form (such as the dpKdV and dLV equations), there is a natural splitting of [**u**] into $\mathcal{I}$ and [$\mathcal{A}$]. By definition, each conservation law $\mathcal{C}(\mathbf{n}, \mathcal{I}, [\mathcal{A}])$ satisfies

$$\mathcal{C}(\mathbf{n}, \mathcal{I}, [0]) \equiv 0. \tag{6.38}$$

As $\mathcal{A}$ has a special role, we will regard it as the *principal variable*, with $(\mathbf{n}, \mathcal{I})$ as *subsidiary* variables. This type of splitting is useful, as a consequence of the following lemma.

Lemma 6.9 *Let* $\mathbf{v} = (v^1, \ldots, v^K)$ *be continuous principal variables and let* $\mathbf{q}$ *be any tuple of subsidiary variables; it is assumed that all variables depend on* $\mathbf{n} \in \mathbb{Z}^N$. *Suppose that* $f = f(\mathbf{q}, [\mathbf{v}])$ *depends on a finite number of shifted principal variables, that it is differentiable with respect to each such* $v_{\mathbf{J}}^\alpha$ *and that every finite shift of* $\partial f / \partial v_{\mathbf{J}}^\alpha$ *is finite. If* $\mathbf{g} = (g^1, \ldots, g^K)$ *is finite on* $\mathbb{Z}^N$, *there exist functions* $h^1, \ldots, h^N$ *such that*

$$(S_{\mathbf{J}} g^\alpha) \frac{\partial f}{\partial v_{\mathbf{J}}^\alpha} = g^\alpha S_{-\mathbf{J}} \left(\frac{\partial f}{\partial v_{\mathbf{J}}^\alpha} \right) + \sum_{i=1}^N (S_i - I) h^i. \tag{6.39}$$

(The summation convention applies to the indices α *and* $\mathbf{J}$ *above.) Moreover, each* h^i *is linear in* [$\mathbf{g}$].

Proof First, consider a single α and $\mathbf{J}$. Define the restricted shift operator $S_{\mathbf{J}}^{\hat{1}} = S_2^{j_2} \cdots S_N^{j_N}$, so that $S_{\mathbf{J}} = S_1^{j_1} S_{\mathbf{J}}^{\hat{1}}$. Elementary summation by parts gives

$$(S_{\mathbf{J}} g^\alpha) \frac{\partial f}{\partial v_{\mathbf{J}}^\alpha} = \left(S_{\mathbf{J}}^{\hat{1}} g^\alpha \right) S_1^{-j_1} \left(\frac{\partial f}{\partial v_{\mathbf{J}}^\alpha} \right) + (S_1 - I) \sigma_k \left\{ \left(S_1^k S_{\mathbf{J}}^{\hat{1}} g^\alpha \right) S_1^{k-j_1} \left(\frac{\partial f}{\partial v_{\mathbf{J}}^\alpha} \right); 0, j_1 \right\}.$$

The proof is completed by iteration, so that $S_2, \ldots, S_N$ are moved in turn, then summing the result over α and $\mathbf{J}$. The sums converge, because each has a finite number of (finite) terms; linearity is obvious.   □

Lemma 6.9 introduces a linear operator, called the (difference) *Euler operator* with respect to **v**, which is

$$\mathbf{E_v} = (E_{v^1}, \ldots, E_{v^K}), \quad \text{where} \quad E_{v^\alpha} = S_{-\mathbf{J}} \frac{\partial}{\partial v_{\mathbf{J}}^\alpha}. \tag{6.40}$$

Thus (6.39) amounts to

$$(S_{\mathbf{J}} g^\alpha) \frac{\partial f}{\partial v_{\mathbf{J}}^\alpha} = g^\alpha E_{v^\alpha}(f) + \sum_{i=1}^N (S_i - I) h^i. \tag{6.41}$$

The Euler operator has a useful alternative form, which is valid if no v^α can be written in terms of $[\mathbf{q}]$. Whenever this condition is satisfied,

$$\mathbf{E}_{v^\alpha}\big(f(\mathbf{q},[\mathbf{v}])\big) = \frac{\partial}{\partial v^\alpha}\left(\sum_{\mathbf{J}} S_{-\mathbf{J}}\big(f(\mathbf{q},[\mathbf{v}])\big)\right). \tag{6.42}$$

Let $\mathbf{v} \mapsto \mathbf{w}$ be a point transformation, and suppose that (6.42) is valid. By applying the chain rule to (6.42), one obtains the change-of-variables formula

$$\mathbf{E}_{v^\alpha} = \frac{\partial w^\beta}{\partial v^\alpha}\,\mathbf{E}_{w^\beta}\,. \tag{6.43}$$

However, this formula does not apply if one or more of the principal variables v^α can be written in terms of $[\mathbf{q}]$.

Supplementary note: The conditions in Lemma 6.9 are far more stringent than is necessary. It is enough that, for each $v_{\mathbf{J}}^\alpha$ that occurs in f, there is a path from $\mathbf{J}$ to $\mathbf{0}$ along which one can sum by parts, a step at a time. Here, a *path* is an ordered set of points in which every point is adjacent to its immediate neighbours.

From here on, we will assume that $f = f(\mathbf{q},[\mathbf{v}])$ meets the conditions of Lemma 6.9 (or its supplementary note) whenever (6.41) is used; these will be called the *summation conditions*.

Corollary 6.10   *Let* $\mathbf{v}(t) = t\mathbf{v} + (1-t)\mathbf{a}(\mathbf{q})$, *where* $\mathbf{a}$ *is chosen to ensure that* $f_t = f(\mathbf{q},[\mathbf{v}(t)])$ *meets the summation conditions for all* $t \in (0,1)$. *Then there exist functions* h^i, *which are identically zero when* $\mathbf{v} = \mathbf{a}(\mathbf{q})$, *such that*

$$f(\mathbf{q},[\mathbf{v}]) = f(\mathbf{q},[\mathbf{a}(\mathbf{q})]) + (v^\alpha - a^\alpha(\mathbf{q}))\int_{t=0}^{1} \mathbf{E}_{v^\alpha(t)}(f_t)\,dt + \sum_{i=1}^{N}(S_i - I)h^i([\mathbf{q}],[\mathbf{v}]). \tag{6.44}$$

Consequently, if $\mathbf{E}_{\mathbf{v}}(f) = 0$ *then* $f(\mathbf{q},[\mathbf{v}]) - f(\mathbf{q},[\mathbf{a}(\mathbf{q})])$ *is a divergence.*

Proof   By Lemma 6.9, there exist functions h_t^i such that

$$\frac{df_t}{dt} = S_{\mathbf{J}}\,(v^\alpha - a^\alpha(\mathbf{q}))\,\frac{\partial f_t}{\partial v_{\mathbf{J}}^\alpha(t)}$$

$$= (v^\alpha - a^\alpha(\mathbf{q}))\,\mathbf{E}_{v^\alpha(t)}(f_t) + \sum_{i=1}^{N}(S_i - I)h_t^i.$$

Every term in h_t^i is the product of a finite function of $([\mathbf{q}],[\mathbf{v}])$ and a shift of a component of $\mathbf{v} - \mathbf{a}(\mathbf{q})$. Integrate over t to obtain (6.44). $\square$

It is usual to choose $\mathbf{a}(\mathbf{q}) = \mathbf{0}$, unless this makes the integral in (6.44) singular. Then, for any system of M PΔEs that can be put into Kovalevskaya form,

Corollary 6.10 gives a reformulation of each conservation law:

$$C(\mathbf{n}, \mathcal{I}, [\mathcal{A}]) = \mathcal{A}_\alpha \int_{t=0}^1 \mathbf{E}_{t\mathcal{A}_\alpha}(C(\mathbf{n}, \mathcal{I}, [t\mathcal{A}]))\, dt + \sum_{i=1}^N (\mathbf{S}_i - \mathbf{I}) h^i(\mathbf{n}, \mathcal{I}, [\mathcal{A}]), \quad (6.45)$$

which follows from (6.38); here $[\mathcal{I}]$ is written in terms of $\mathbf{n}$, $\mathcal{I}$ and $[\mathcal{A}]$. Furthermore, $h^i(\mathbf{n}, \mathcal{I}, [0]) \equiv 0$, so the conservation law (6.45) is equivalent to

$$\overline{C}(\mathbf{n}, \mathcal{I}, [\mathcal{A}]) = \mathcal{A}_\alpha \int_{t=0}^1 \mathbf{E}_{t\mathcal{A}_\alpha}(C(\mathbf{n}, \mathcal{I}, [t\mathcal{A}]))\, dt, \quad (6.46)$$

which is in characteristic form. The characteristic is $\mathcal{Q} = (\mathcal{Q}^1, \dots, \mathcal{Q}^M)$, where

$$\mathcal{Q}^\alpha(\mathbf{n}, \mathcal{I}, [\mathcal{A}]) = \int_{t=0}^1 \mathbf{E}_{t\mathcal{A}_\alpha}(C(\mathbf{n}, \mathcal{I}, [t\mathcal{A}]))\, dt. \quad (6.47)$$

Shortly, we will prove that the value of $\mathcal{Q}$ on solutions of the PΔE identifies the conservation law C uniquely, up to equivalence. Consequently, we define the *root* of this conservation law to be $\overline{\mathcal{Q}} = (\overline{\mathcal{Q}}^1, \dots, \overline{\mathcal{Q}}^M)$, where

$$\overline{\mathcal{Q}}^\alpha(\mathbf{n}, \mathcal{I}) = \mathcal{Q}^\alpha(\mathbf{n}, \mathcal{I}, [0]) = \{\mathbf{E}_{\mathcal{A}_\alpha}(C(\mathbf{n}, \mathcal{I}, [\mathcal{A}]))\}|_{[\mathcal{A}=0]}. \quad (6.48)$$

(The second equality holds if C is locally analytic (in $[\mathcal{A}]$) at $\mathcal{A} = 0$; this condition is satisfied except in pathological cases.) It is quite straightforward to calculate the root of a conservation law, as the following example shows.

Example 6.11 The dLV equation (6.34) has conservation laws that include

$$C_1(m, n, \mathcal{I}, [\mathcal{A}]) = (u_{1,0} - 1)\mathcal{A},$$
$$C_2(m, n, \mathcal{I}, [\mathcal{A}]) = (\mathcal{A} + \omega)\, \mathbf{S}_n \mathcal{A} + (u_{1,0} - 2)\mathcal{A}.$$

From (6.48), the root of C_2 is

$$\overline{\mathcal{Q}}(m, n, \mathcal{I}) = \{\mathbf{S}_n \mathcal{A} + \mathbf{S}_n^{-1}(\mathcal{A} + \omega) + u_{1,0} - 2\}|_{[\mathcal{A}=0]} = 2(u_{1,0} - 1),$$

which coincides with the root of $2C_1$, as expected. The dLV equation also has the conservation law whose components are

$$(F_3, G_3) = \left((-1)^{m+n} \ln\{u(u_{0,1} - 1)\},\; (-1)^{m+n+1} \ln\{u_{1,0}(u_{1,0} - 1)\}\right).$$

This conservation law amounts to

$$C_3(m, n, \mathcal{I}, [\mathcal{A}]) = (-1)^{m+n} \ln\left\{\frac{(u_{1,0} - 1)\mathcal{A} + u(u_{0,1} - 1)}{u(u_{0,1} - 1)}\right\},$$

so its root is

$$\overline{\mathcal{Q}}(m, n, \mathcal{I}) = \frac{(-1)^{m+n}(u_{1,0} - 1)}{u(u_{0,1} - 1)}. \qquad \blacktriangle$$

Having seen how the root is calculated, we now establish the importance of the root for a given PΔE in Kovalevskaya form,

$$\mathcal{A}_\alpha \equiv u^\alpha_{p,\mathbf{K}_\alpha} - \omega^\alpha(\mathbf{n}, \mathcal{I}) = 0, \qquad \alpha = 1, \ldots, M, \tag{6.49}$$

where $\mathcal{I} = \left(\mathbf{u}_{\{0,\cdot\}}, \mathbf{u}_{\{1,\cdot\}}, \ldots, \mathbf{u}_{\{p-1,\cdot\}}\right)$. (As usual, we restrict attention to PΔEs that are locally smooth in their continuous arguments.)

Theorem 6.12 *For any PΔE that is in Kovalevskaya form, the conservation law $\mathcal{C}(\mathbf{n}, \mathcal{I}, [\mathcal{A}])$ is trivial if and only if its root is zero. Consequently, two conservation laws are equivalent if and only if they have the same root.*

Proof By construction, the root depends linearly on $\mathcal{C}$, so the second statement follows immediately from the first. Only the first kind of triviality can affect $\mathcal{C}$ (and hence, the root). So, if $\mathcal{C}$ is trivial, one can assume without loss of generality that $F^i = F^i(\mathbf{n}, \mathcal{I}, [\mathcal{A}])$, where $F^i(\mathbf{n}, \mathcal{I}, [0]) \equiv 0$. Then the root has components

$$\overline{\mathcal{Q}}^\alpha(\mathbf{n}, \mathcal{I}) = \left\{ \mathbf{E}_{\mathcal{A}_\alpha} \left(\sum_{i=1}^N (\mathbf{S}_i - 1)\left(F^i(\mathbf{n}, \mathcal{I}, [\mathcal{A}]) - F^i(\mathbf{n}, \mathcal{I}, [0]) \right) \right) \right\} \Bigg|_{[\mathcal{A}=0]}$$

$$= \sum_{i=1}^N \left\{ \mathbf{S}_{-\mathbf{J}} \left(\mathbf{S}_i \left(\frac{\partial F^i(\mathbf{n}, \mathcal{I}, [\mathcal{A}])}{\partial \mathbf{S}_{\mathbf{J}-1_i} \mathcal{A}_\alpha} \right) - \frac{\partial F^i(\mathbf{n}, \mathcal{I}, [\mathcal{A}])}{\partial \mathbf{S}_{\mathbf{J}} \mathcal{A}_\alpha} \right) \right\} \Bigg|_{[\mathcal{A}=0]} = 0.$$

To prove the converse, suppose that the root is zero. If the components of a conservation law with this root depend on $[\mathcal{A}]$ then $F^i(\mathbf{n}, \mathcal{I}, [\mathcal{A}]) - F^i(\mathbf{n}, \mathcal{I}, [0])$ are the components of a trivial conservation law, which can be subtracted without affecting the root. So, without loss of generality, assume that $F^i = F^i(\mathbf{n}, \mathcal{I})$. The PΔE (6.49) appears in only one part of the conservation law, namely $\mathbf{S}_1 F^1$. Therefore

$$0 \equiv \overline{\mathcal{Q}}^\alpha(\mathbf{n}, \mathcal{I}) = \left\{ \mathbf{S}^{\hat{1}}_{-\mathbf{J}} \left(\frac{\partial \mathbf{S}_1 F^1}{\partial \mathbf{S}^{\hat{1}}_{\mathbf{J}} u^\alpha_{p, \mathbf{K}_\alpha}} \right) \right\} \Bigg|_{\mathcal{I}} = \mathbf{E}^{\hat{1}}_{\omega^\alpha} \left(\left\{ \mathbf{S}_1 F^1 \right\} \big|_{\mathcal{I}} \right); \tag{6.50}$$

here $\mathbf{S}^{\hat{1}}_{\mathbf{J}}$ is the restricted shift operator, $\mathbf{S}^{\hat{1}}_{-\mathbf{J}}$ is its inverse, and $\mathbf{E}^{\hat{1}}_{\omega^\alpha}$ is a restricted Euler operator that treats ω^α as the principal variable and n^1 as a parameter. Furthermore, the right-hand side of (6.50) involves $\mathbf{u}_{\{0,\cdot\}}$ only through ω and its shifts. By Corollary 6.10 (with $a = 0$, and with the subsidiary variables $n^2, \ldots, n^N, \mathbf{u}_{\{1,\cdot\}}, \ldots, \mathbf{u}_{\{p-1,\cdot\}}$), there exist h^i such that

$$\left\{ \mathbf{S}_1 F^1 \right\} \big|_{\mathcal{I}} = \mathbf{S}_1 \left\{ F^1 \left(\mathbf{n}, \mathbf{u}_{\{0,\cdot\}}, \ldots, \mathbf{u}_{\{p-2,\cdot\}}, \mathbf{0} \right) \right\} + \sum_{i=2}^N (\mathbf{S}_i - 1) \left\{ \mathbf{S}_1 h^i(\mathbf{n}, \mathcal{I}) \right\} \big|_{\mathcal{I}},$$

and so, up to a trivial component,

$$F^1 = F^1\left(\mathbf{n}, \mathbf{u}_{\{0,\cdot\}}, \dots, \mathbf{u}_{\{p-2,\cdot\}}, \mathbf{0}\right) + \sum_{i=2}^{N}(S_i - I)\,h^i(\mathbf{n}, \mathcal{I}).$$

This is equivalent to a conservation law of the same form, but with $h^i = 0$. Such a conservation law is necessarily trivial, because it does not involve the PΔE (6.49) anywhere.   □

Note: Although the proof of Theorem 6.12 uses Kovalevskaya form for simplicity, the theorem holds in any coordinate system. (If an expression is a conservation law (or, more generally, a divergence), it remains so under any valid change of lattice coordinates, and triviality is likewise preserved; the proof of this is left as an easy exercise.) However, the root is coordinate-dependent. It will change if one uses different principal variables, $\tilde{\mathcal{A}}_\alpha = \theta_\alpha(\mathbf{n}, \mathcal{I}, \mathcal{A})$, where $\tilde{\mathcal{A}} = 0$ for all $(\mathbf{n}, \mathcal{I})$ if and only if $\mathcal{A} = 0$, and where both of these forms of the PΔE satisfy the maximal rank condition. The effect of coordinate changes on the root is examined in Exercise 6.9.

Sometimes one can use generalized Lie symmetries to generate new conservation laws from a known conservation law. The infinitesimal generator X commutes with each shift operator, so if $\mathcal{C}$ is a conservation law with components $\mathbf{F}$ then $\widetilde{\mathcal{C}} = X\mathcal{C}$ is a divergence expression with components $\widetilde{\mathbf{F}} = X\mathbf{F}$. Clearly, if $\mathcal{C}$ is a trivial conservation law, so is $\widetilde{\mathcal{C}}$. (Remember that $X\mathcal{A}_\alpha = 0$ when $[\mathcal{A} = 0]$.) Thus, if the PΔE can be written in Kovalevskaya form, we may assume that $\mathcal{C}$ is in characteristic form:

$$\mathcal{C}(\mathbf{n}, \mathcal{I}, [\mathcal{A}]) = \mathcal{A}_\alpha \mathcal{Q}^\alpha(\mathbf{n}, \mathcal{I}, [\mathcal{A}]). \tag{6.51}$$

Then $\widetilde{\mathcal{C}}$ is a conservation law, because

$$\widetilde{\mathcal{C}}(\mathbf{n}, \mathcal{I}, [\mathcal{A}]) = X(\mathcal{A}_\alpha)\,\mathcal{Q}^\alpha(\mathbf{n}, \mathcal{I}, [\mathcal{A}]) + \mathcal{A}_\alpha\,X(\mathcal{Q}^\alpha(\mathbf{n}, \mathcal{I}, [\mathcal{A}]))$$

satisfies $\widetilde{\mathcal{C}}(\mathbf{n}, \mathcal{I}, [0]) \equiv 0$. Of course, there is no guarantee that this conservation law is genuinely new, or even nontrivial! However, one can find this out easily by calculating the root of $\widetilde{\mathcal{C}}$ (preferably, using a computer algebra system).

Example 6.13   The generalized symmetry generators for the dLV equation,

$$\mathcal{A} \equiv u_{1,1} - \omega = 0, \qquad \omega = \frac{u\,(u_{0,1} - 1)}{u_{1,0} - 1},$$

include $X = u(u - 1)(u_{0,1} - u_{0,-1})\,\partial_u$. Applying X to the conservation law

$\mathcal{C} = (u_{1,0} - 1)\mathcal{A}$, we obtain an apparently new conservation law:

$$
\begin{aligned}
\widetilde{\mathcal{C}} &= u_{1,1}(u_{1,0} - 1)\{u_{1,2}(u_{1,1} - 1) - u_{1,0}(u_{1,-1} - 1)\} \\
&\quad + u(u_{0,1} - 1)\{u_{0,-1}(u - 1) - u_{0,1}(u_{0,2} - 1)\} \\
&= (u_{1,0} - 1)(\mathcal{A} + \omega)(\mathcal{A} + \omega - 1)\,\mathrm{S}_n\mathcal{A} + u_{0,1}(u_{0,2} - 1)(u_{1,0} - 1)\mathcal{A} \\
&\quad + \frac{u_{0,-1}(u - 1)(u_{1,0} - 1)\left(u_{1,0}\,\mathcal{A} + \omega\,\mathrm{S}_n^{-1}\mathcal{A}\right)}{\mathrm{S}_n^{-1}\mathcal{A} - u_{1,0}}.
\end{aligned}
$$

In this case, appearances are deceptive; after a straightforward (but slightly messy) calculation, it turns out that the root of $\widetilde{\mathcal{C}}$ is zero, so the 'new' conservation law is trivial. This shows the benefit of having a systematic test for equivalence. ▲

6.5 Variational PΔEs

The rest of this chapter deals with PΔEs that stem from a variational problem. We have already met most of the basic ideas; readers may find it helpful to revisit §3.7 and §3.8 before continuing. Once again, it is helpful to work formally, assuming that any criteria that are necessary for convergence are satisfied, as are local smoothness conditions. We will also make the blanket assumption that the summation conditions in Lemma 6.9 or its supplementary note are satisfied (typically, with **u** as the principal variables and **n** as the subsidiary variables) whenever it is necessary to sum by parts.

The basic variational problem is to find extrema of a given functional,

$$
\mathscr{L}[\mathbf{u}] = \sum L(\mathbf{n}, [\mathbf{u}]), \tag{6.52}
$$

where the sum is over a regular domain and **u** is prescribed at the boundaries. Extrema are found just as for variational OΔE problems, by requiring that

$$
\left\{\frac{\mathrm{d}}{\mathrm{d}\epsilon}\,\mathscr{L}[\mathbf{u} + \epsilon\boldsymbol{\eta}]\right\}\bigg|_{\epsilon=0} = 0, \tag{6.53}
$$

for all $\boldsymbol{\eta}$ that vanish on the boundary (or in the appropriate limit, where the domain is unbounded). When (6.52) is summed by parts, these boundary constraints eliminate contributions from divergence terms, so (6.52) amounts to

$$
0 = \sum \left(\mathrm{S}_{\mathbf{J}}\eta_{\mathbf{J}}^{\alpha}\right)\frac{\partial L}{\partial u_{\mathbf{J}}^{\alpha}} = \sum \eta^{\alpha}\,\mathrm{S}_{-\mathbf{J}}\left(\frac{\partial L}{\partial u_{\mathbf{J}}^{\alpha}}\right) = \sum \eta^{\alpha}\,\mathbf{E}_{u^{\alpha}}(L).
$$

This holds for arbitrary $\boldsymbol{\eta}$ only if **u** satisfies the Euler–Lagrange equations:

$$
\mathbf{E}_{u^{\alpha}}(L) = 0, \qquad \alpha = 1, \dots, M. \tag{6.54}
$$

In general, it is easier to calculate Euler–Lagrange equations for PΔEs than for PDEs, as only one differentiation is needed for each term[4].

Example 6.14   For the Lagrangian

$$L = \ln \left| \frac{u_{1,0} - u_{0,1}}{u_{1,1} - u} \right|, \tag{6.55}$$

the Euler–Lagrange equation is

$$\frac{\partial L}{\partial u} + S_m^{-1}\left(\frac{\partial L}{\partial u_{1,0}}\right) + S_n^{-1}\left(\frac{\partial L}{\partial u_{0,1}}\right) + S_m^{-1}S_n^{-1}\left(\frac{\partial L}{\partial u_{1,1}}\right) = 0,$$

which amounts to

$$\mathbf{E}_u(L) \equiv \frac{1}{u_{1,1} - u} + \frac{1}{u - u_{-1,1}} - \frac{1}{u_{1,-1} - u} - \frac{1}{u - u_{-1,-1}} = 0. \tag{6.56}$$

This is a Toda-type equation that is satisfied by all solutions of the autonomous dpKdV equation (among others).   ▲

Example 6.15   As a simple non-scalar example, consider the Lagrangian

$$L = \frac{u_{1,0} - v}{u - v_{0,1}}, \tag{6.57}$$

which yields the following system of Euler–Lagrange equations:

$$\mathbf{E}_u(L) \equiv \frac{1}{u_{-1,0} - v_{-1,1}} - \frac{u_{1,0} - v}{(u - v_{0,1})^2} = 0,$$
$$\mathbf{E}_v(L) \equiv -\frac{1}{u - v_{0,1}} + \frac{u_{1,-1} - v_{0,-1}}{(u_{0,-1} - v)^2} = 0. \tag{6.58}$$

Although (6.58) looks complicated, a little manipulation (which is left as an exercise) yields the general solution

$$u = \left(\frac{f^2}{f_{-1}} - f_1\right)m + g, \qquad v = \left(\frac{f_1^2}{f} - f_2\right)m - f_2 + g_1; \tag{6.59}$$

here $f = f(m - n)$ and $g = g(m - n)$ are arbitrary, except that f is nonzero.   ▲

Any function $L(\mathbf{n}, [\mathbf{u}])$ that satisfies $\mathbf{E}_{u^\alpha}(L) \equiv \mathbf{0}$ for each α is said to be a *null Lagrangian*. The following theorem characterizes such functions.

Theorem 6.16   *A function $L(\mathbf{n}, [\mathbf{u}])$ that is defined on a regular domain is a null Lagrangian if and only if it is a divergence.*

---

[4] However, there exist Lagrangians whose Euler–Lagrange equations have no solutions (see Exercise 6.12).

Proof   If $\mathbf{E}_{u^\alpha}(L) \equiv 0$ then, by Corollary 6.10 with $\mathbf{v} = \mathbf{u}$, $\mathbf{q} = \mathbf{n}$ and $\mathbf{a} = \mathbf{0}$,

$$L(\mathbf{n}, [\mathbf{u}]) = L(\mathbf{n}, [\mathbf{0}]) + \sum_{i=1}^{N}(\mathrm{S}_i - \mathrm{I})h^i(\mathbf{n}, [\mathbf{u}]).$$

Choose lattice coordinates $\mathbf{n} = (n^1, n^2, \ldots, n^N)$ such that the domain is a box that contains the point $\mathbf{0}$. (As was noted earlier, a divergence remains a divergence under a valid change of lattice coordinates.) Any function $f(\mathbf{n})$ that is defined on a box may be written as a divergence, as follows. First, observe that

$$f(n^1, n^2, \ldots, n^N) = f(0, n^2, \ldots, n^N) + (\mathrm{S}_1 - \mathrm{I})\,\sigma_k\left\{f(k, n^2, \ldots, n^N); 0, n^1\right\}.$$

Iterate to get an expression of the form

$$f(n^1, n^2, \ldots, n^N) = f(\mathbf{0}) + \sum_{i=1}^{N}(\mathrm{S}_i - \mathrm{I})\tilde{h}^i(\mathbf{n}).$$

If the final term, $f(\mathbf{0})$, is nonzero, incorporate it into the divergence by adding $f(\mathbf{0})\,n^1$ to $\tilde{h}^1$. So $L(\mathbf{n}, [\mathbf{0}])$ is a divergence, and thus, so is every null Lagrangian. To prove the converse, note that for every differentiable function $f^i(\mathbf{n}, [\mathbf{u}])$,

$$\mathbf{E}_{u^\alpha}\left((\mathrm{S}_i - \mathrm{I})(f^i)\right) = \mathrm{S}_{-\mathbf{J}}\left(\mathrm{S}_i\left(\frac{\partial f^i}{\partial u^\alpha_{\mathbf{J}-\mathbf{1}_i}}\right)\right) - \mathbf{E}_{u^\alpha}(f^i) \equiv 0. \tag{6.60}$$

All that remains is to sum this result over i.   □

The freedom to add null Lagrangians is used from here on to ensure that L contains no $\mathbf{u}_{\mathbf{J}}$ for which any j^i is negative.

6.6 Noether's Theorem for PΔEs

Noether's Theorem for PΔEs is derived similarly to the corresponding result for OΔEs. However, dynamical symmetries are replaced by generalized symmetries. Each symmetry generator,

$$X = \mathrm{S}_{\mathbf{J}}(Q^\alpha(\mathbf{n}, [\mathbf{u}]))\frac{\partial}{\partial u^\alpha_{\mathbf{J}}},$$

for the Euler–Lagrange equations satisfies the LSC:

$$X(\mathbf{E}_{u^\alpha}(L)) = 0 \quad \text{when (6.54) holds,} \qquad \alpha = 1, \ldots, M.$$

The generalized symmetries generated by X are *variational* if they preserve the Euler–Lagrange equations exactly, that is, if there exist functions $P^i_0(\mathbf{n}, [\mathbf{u}])$

such that

$$X(L) \equiv \sum_{i=1}^{N} (S_i - I) P_0^i(\mathbf{n}, [\mathbf{u}]). \tag{6.61}$$

Clearly, the set of variational symmetry generators is a vector space. Let P_0^i (resp. $\tilde{P}_0^i$) be the functions in (6.61) that correspond to X (resp. $\tilde{X}$). Then

$$[X, \tilde{X}](L) = \sum_{i=1}^{N} (S_i - I) \left\{ X \left(\tilde{P}_0^i(\mathbf{n}, [\mathbf{u}]) \right) - \tilde{X} \left(P_0^i(\mathbf{n}, [\mathbf{u}]) \right) \right\},$$

and so the vector space of variational symmetry generators is a Lie algebra.

By now, we have enough machinery to prove the difference analogues of Noether's two theorems on variational symmetries. (Reminder: we are assuming that summation by parts is valid; otherwise, these theorems will not hold.)

Theorem 6.17 (Noether's Theorem for PΔEs)   *A prolonged vector field X is a variational symmetry generator for the functional* (6.52) *if and only if its characteristic $\mathbf{Q}$ is the characteristic of a conservation law for the Euler–Lagrange equations.*

Proof   Suppose that the prolonged vector field X generates variational symmetries. Summing (6.61) by parts (using the method in the proof of Lemma 6.9), we obtain

$$X(L) = Q^\alpha \mathbf{E}_{u^\alpha}(L) + \sum_{i=1}^{N} (S_i - I) h^i(\mathbf{n}, [\mathbf{u}]). \tag{6.62}$$

This yields a conservation law in characteristic form:

$$\mathcal{C} \equiv Q^\alpha \mathbf{E}_{u^\alpha}(L) = \sum_{i=1}^{N} (S_i - I) P^i(\mathbf{n}, [\mathbf{u}]), \qquad P^i = P_0^i - h^i, \tag{6.63}$$

whose characteristic is $\mathbf{Q}$. To prove the converse, start with (6.63) and sum by parts to show that $X(L)$ is a divergence, where $X = (S_J Q^\alpha) \partial/\partial u_J^\alpha$.   □

Example 6.18   The Toda-type equation (6.56) admits scaling symmetries whose characteristic is $Q = u$. It is straightforward to check that these are variational symmetries for the Lagrangian (6.55) and that $P_0^1 = P_0^2 = 0$. The corresponding conservation law is

$$(S_m - I) \left(\frac{u}{u - u_{-1,1}} - \frac{u}{u - u_{-1,-1}} \right) + (S_n - I) \left(\frac{u}{u - u_{1,-1}} - \frac{u_{1,0}}{u_{1,0} - u_{0,-1}} \right) = 0. \tag{6.64}$$

As (6.56) is satisfied by all solutions of the autonomous dpKdV equation, (6.64) is also a conservation law for that equation, whose characteristics do not include $Q = u$. This result does not clash with Noether's Theorem, because the autonomous dpKdV equation is not an Euler–Lagrange equation. It turns out that the conservation law is trivial for the autonomous dpKdV equation, even though it is not trivial for the Toda-type equation (6.56). ▲

Example 6.19 The system of Euler–Lagrange equations (6.58) has (among others) the variational symmetry generator

$$X = u\partial_u + v\partial_v.$$

Applying this to the Lagrangian (6.57) gives $XL = 0$, so Noether's Theorem yields the conservation law

$$C = u\,\mathbf{E}_u(L) + v\,\mathbf{E}_v(L)$$

$$= u\left(\frac{1}{u_{-1,0} - v_{-1,1}} - \frac{u_{1,0} - v}{(u - v_{0,1})^2}\right) + v\left(-\frac{1}{u - v_{0,1}} + \frac{u_{1,-1} - v_{0,-1}}{(u_{0,-1} - v)^2}\right)$$

$$= (S_m - I)\left(-\frac{u}{u_{-1,0} - v_{-1,1}}\right) + (S_n - I)\left(-\frac{v(u_{1,-1} - v_{0,-1})}{(u_{0,-1} - v)^2}\right). \quad\blacktriangle$$

Noether's Theorem establishes the link between variational symmetries and conservation laws. It leads to a stronger result for Euler–Lagrange equations that can be written in Kovalevskaya form.

Theorem 6.20 *Suppose that the Euler–Lagrange equations can be written in Kovalevskaya form. Then there is a one-to-one correspondence between the equivalence classes of variational symmetries and the equivalence classes of conservation laws of the Euler–Lagrange equations.*

Proof If C and $\widetilde{C}$ are two equivalent conservation laws, Theorem 6.12 states that they have the same root, $\overline{Q}$ (with respect to a given set of initial variables, $\mathcal{I}$). By Noether's Theorem, their characteristics, Q and $\widetilde{Q}$, are also characteristics of variational symmetries; these are equivalent to one another, because they share the same basic characteristic (with respect to $\mathcal{I}$), which is $\overline{Q}$.

Conversely, suppose that $\mathbf{Q}$ and $\widetilde{\mathbf{Q}}$ are equivalent characteristics of variational symmetries, so that both are equivalent to the same basic characteristic, $\mathbf{Q}_{\mathcal{I}}$. By Noether's Theorem, $\mathbf{Q}$ and $\widetilde{\mathbf{Q}}$ are characteristics of conservation laws. These are equivalent, as they share a common root (namely $\mathbf{Q}_{\mathcal{I}}$). □

Note: In general, a root need not be a characteristic of a conservation law. Similarly, a basic characteristic need not be a variational symmetry characteristic. It is not yet known whether every basic characteristic for a given Euler–Lagrange equation is equivalent to a variational symmetry characteristic.

6.7 Noether's Second Theorem

The analogue of Noether's Second Theorem for the difference calculus of variations is as follows.

Theorem 6.21 *The variational problem* (6.53) *admits an infinite-dimensional Lie algebra of generalized variational symmetry generators, whose characteristics* $\mathbf{Q}(\mathbf{n}, [\mathbf{u}; \mathbf{g}])$ *depend on* R *arbitrary functions* $\mathbf{g} = (g^1(\mathbf{n}), \ldots, g^R(\mathbf{n}))$ *and their shifts, if and only if there exist linear difference operators* $\mathcal{D}_r^\alpha$ *that yield* R *independent difference relations between the Euler–Lagrange equations:*

$$\mathcal{D}_r^\alpha \mathbf{E}_{u^\alpha}(L) \equiv 0, \qquad r = 1, \ldots, R. \qquad (6.65)$$

Proof Suppose that there exists a set of difference relations (6.65). Multiply the rth such relation by the arbitrary function $g^r(\mathbf{n})$, sum over r, and sum by parts to get a trivial conservation law of the form

$$Q^\alpha(\mathbf{n}, [\mathbf{u}; \mathbf{g}])\, \mathbf{E}_{u^\alpha}(L) = \sum_{i=1}^{N}(S_i - I)P^i(\mathbf{n}, [\mathbf{u}; \mathbf{g}]), \qquad (6.66)$$

where

$$Q^\alpha(\mathbf{n}, [\mathbf{u}; \mathbf{g}]) = (\mathcal{D}_r^\alpha)^\dagger(g^r).$$

By Noether's Theorem, $\mathbf{Q}$ is a characteristic of a variational symmetry generator (for every choice of the arbitrary functions g^r).

To prove the converse, the key is to regard each g^r as a principal variable and $(\mathbf{n}, [\mathbf{u}])$ as subsidiary variables. Applying the Euler–Lagrange operator $\mathbf{E}_{g^r}$ to (6.66) gives

$$\mathbf{E}_{g^r}\{Q^\alpha(\mathbf{n}, [\mathbf{u}; \mathbf{g}])\, \mathbf{E}_{u^\alpha}(L)\} \equiv S_{-\mathbf{J}}\left(\frac{\partial Q^\alpha(\mathbf{n}, [\mathbf{u}; \mathbf{g}])}{\partial g_{\mathbf{J}}^r}\, \mathbf{E}_{u^\alpha}(L)\right) = 0. \qquad (6.67)$$

The final equality in (6.67) is obtained by the same reasoning that leads to (6.60), bearing in mind that no g^r can be written in terms of $(\mathbf{n}, [\mathbf{u}])$. Consequently, the required difference operators are

$$\mathcal{D}_r^\alpha = \sum_{\mathbf{J}}\left\{S_{-\mathbf{J}}\left(\frac{\partial Q^\alpha(\mathbf{n}, [\mathbf{u}; \mathbf{g}])}{\partial g_{\mathbf{J}}^r}\right)\right\} S_{-\mathbf{J}}.$$

It is convenient to regard (6.67) as 'new' Euler–Lagrange equations that arise when $\mathbf{g}$ is varied in the functional

$$\hat{\mathscr{L}}[\mathbf{u}; \mathbf{g}] = \sum \hat{L}(\mathbf{n}, [\mathbf{u}; \mathbf{g}]), \qquad (6.68)$$

where

$$\hat{L}(\mathbf{n}, [\mathbf{u}; \mathbf{g}]) = Q^\alpha(\mathbf{n}, [\mathbf{u}; \mathbf{g}])\, \mathbf{E}_{u^\alpha}(L(\mathbf{n}, [\mathbf{u}])). \qquad (6.69)$$

To prove that the relations (6.67) are independent, suppose that the converse is true; then there exist linear difference operators $\hat{\mathcal{D}}^r$ such that

$$\hat{\mathcal{D}}^1 \mathbf{E}_{g^1}(\hat{L}) + \cdots + \hat{\mathcal{D}}^R \mathbf{E}_{g^R}(\hat{L}) \equiv 0.$$

By the 'if' part of Noether's Second Theorem, there exist variational symmetries of (6.68) whose generator is of the form

$$\hat{X} = \mathbf{S}_\mathbf{J}\left\{(\hat{\mathcal{D}}^r)^\dagger(\hat{g})\right\}\frac{\partial}{\partial g_\mathbf{J}^r},$$

where $\hat{g}$ is an arbitrary function. That is, there exist functions $\hat{P}^i$ such that

$$\hat{X}\left\{Q^\alpha(\mathbf{n}, [\mathbf{u}; \mathbf{g}])\right\}\mathbf{E}_{u^\alpha}(L(\mathbf{n}, [\mathbf{u}])) = \hat{X}(\hat{L}) = \sum_{i=1}^{N}(\mathbf{S}_i - \mathbf{I})\hat{P}^i(\mathbf{n}, [\mathbf{u}; \mathbf{g}; \hat{g}]).$$

Consequently, for arbitrary $\mathbf{g}$ and $\hat{g}$,

$$\hat{Q}^\alpha(\mathbf{n}, [\mathbf{u}; \mathbf{g}; \hat{g}]) = \hat{X}\left\{Q^\alpha(\mathbf{n}, [\mathbf{u}; \mathbf{g}])\right\}$$

is a characteristic of variational symmetries for the original variational problem (6.53). This implies that the set of variational symmetries for (6.53) depends on more than R independent arbitrary functions, contradicting our original assumption. Hence the relations (6.67) cannot be dependent.

□

Example 6.22 Consider the variational problem whose Lagrangian is

$$L = (v - u_{1,0})(w - u_{0,1}) - \mu(v_{0,1} - w_{1,0})^2,$$

where μ is a nonzero constant. Each variational point symmetry generator that depends on one or more completely arbitrary functions of (m, n) is of the form

$$X = g(m, n)\,\partial_u + g(m + 1, n)\,\partial_v + g(m, n + 1)\,\partial_w;$$

here $XL \equiv 0$. The Euler–Lagrange equations are

$$\mathbf{E}_u(L) \equiv -w_{-1,0} + u_{-1,1} - v_{0,-1} + u_{1,-1} = 0,$$
$$\mathbf{E}_v(L) \equiv w - u_{0,1} - 2\mu(v - w_{1,-1}) = 0,$$
$$\mathbf{E}_w(L) \equiv v - u_{1,0} + 2\mu(v_{-1,1} - w) = 0.$$

As X depends on a single arbitrary function, Noether's Second Theorem gives one difference relation between the Euler–Lagrange equations. So (6.67) is

$$\mathbf{E}_g\left\{g\mathbf{E}_u(L) + g_{1,0}\mathbf{E}_v(L) + g_{1,0}\mathbf{E}_w(L)\right\} \equiv 0,$$

which amounts to the difference relation

$$\mathbf{E}_u(L) + \mathbf{S}_m^{-1}\left(\mathbf{E}_v(L)\right) + \mathbf{S}_n^{-1}\left(\mathbf{E}_w(L)\right) \equiv 0.$$

The three Euler–Lagrange equations are related because the Lagrangian involves only two dependent variables (and their shifts), namely $\tilde{v} = v - u_{1,0}$ and $\tilde{w} = w - u_{0,1}$. If we write

$$L = \tilde{v}\tilde{w} - \mu(\tilde{v}_{0,1} - \tilde{w}_{1,0})^2,$$

the Euler–Lagrange equations corresponding to variations in $\tilde{v}$ and $\tilde{w}$ are

$$\mathbf{E}_{\tilde{v}}(L) \equiv \tilde{w} - 2\mu(\tilde{v} - \tilde{w}_{1,-1}) = 0, \qquad \mathbf{E}_{\tilde{w}}(L) \equiv \tilde{v} + 2\mu(\tilde{v}_{-1,1} - \tilde{w}) = 0.$$

These two equations determine $\tilde{v}$ and $\tilde{w}$ completely, but v and w are determined only up to an arbitrary choice of u. This is an example of gauge freedom, which is expressed by the *gauge symmetry* generator X.   ▲

More generally, variational Lie symmetries whose generator depends on a completely arbitrary function of all independent variables are called gauge symmetries. Much of our current understanding of fundamental physics is based on gauge theories, which can be written as systems of PDEs that incorporate particular gauge symmetries. This creates an important challenge for geometric integration: how can one discretize a gauge theory without destroying gauge symmetries? (See Exercise 6.16 for one instance in which this has been done successfully.)

Usually, the characteristic $\mathbf{Q}$ is linear in $\mathbf{g}$ and its shifts, in which case the relations (6.67) are independent of $\mathbf{g}$. However, with a different choice of $\mathbf{g}$, one could obtain relations that depend on $[\mathbf{g}]$. To resolve this difficulty, apply a point transformation, $\mathbf{g} \mapsto \mathbf{h}$. The change-of-variables formula (6.43) gives

$$\mathbf{E}_{\mathbf{h}^\rho}\{Q^\alpha(\mathbf{n}, [\mathbf{u}; \mathbf{g}(\mathbf{h})]) \, \mathbf{E}_{u^\alpha}(L)\} \equiv \frac{\partial g^r(\mathbf{h})}{\partial \mathbf{h}^\rho} \left(\mathbf{E}_{g^r}\{Q^\alpha(\mathbf{n}, [\mathbf{u}; \mathbf{g}]) \, \mathbf{E}_{u^\alpha}(L)\}\right)\Big|_{\mathbf{g}=\mathbf{g}(\mathbf{h})},$$

so the relations (6.67) are equivalent to

$$\mathbf{E}_{\mathbf{h}^\rho}\{Q^\alpha(\mathbf{n}, [\mathbf{u}; \mathbf{g}(\mathbf{h})]) \, \mathbf{E}_{u^\alpha}(L)\} = 0, \qquad \rho = 1, \ldots, R. \tag{6.70}$$

Provided that there exists a point transformation such that $\mathbf{Q}$ is linear in $[\mathbf{h}]$, the transformed relations (6.70) are independent of any arbitrary functions.

6.8 Constrained variational symmetries

It is common for variational symmetry generators to depend on functions of the independent variables that are constrained in some way. Constraints on these arbitrary functions arise from the linearized symmetry condition for the Euler–Lagrange equation, coupled with the (linear) requirement that the symmetries are variational. Suppose that R functions, $g^r(\mathbf{n})$, are subject to a set of S linear difference constraints,

$$\mathcal{D}_{sr}g^r = 0, \qquad s = 1, \ldots, S, \tag{6.71}$$

and that this set is complete (that is, no additional constraints are needed). The constraints may be applied by taking variations with respect to $\mathbf{g}$ of the augmented Lagrangian

$$\hat{L}(\mathbf{n}, [\mathbf{u}; \mathbf{g}; \lambda]) = Q^\alpha(\mathbf{n}, [\mathbf{u}; \mathbf{g}]) \, \mathbf{E}_{u^\alpha}(L(\mathbf{n}, [\mathbf{u}])) - \lambda^s \mathcal{D}_{sr}(g^r); \tag{6.72}$$

here $\lambda^1, \ldots, \lambda^S$ are Lagrange multipliers. This produces R relations between the Euler–Lagrange equations and the Lagrange multipliers:

$$\mathrm{S}_{-\mathbf{J}}\left(\frac{\partial Q^\alpha(\mathbf{n}, [\mathbf{u}; \mathbf{g}])}{\partial \mathrm{S}_{\mathbf{J}} g^r} \, \mathbf{E}_{u^\alpha}(L)\right) \equiv (\mathcal{D}_{sr})^\dagger (\lambda^s), \qquad r = 1, \ldots, R. \tag{6.73}$$

If $S < R$, one can eliminate the Lagrange multipliers from (6.73) to obtain $R - S$ difference relations between the Euler–Lagrange equations. When the system of constraints (6.71) does not allow the multipliers λ^s to be eliminated, the most that one can do is to construct conservation laws of the Euler–Lagrange equations (6.54), as follows. First, determine a particular solution of (6.73) for λ^s. This is then substituted into the divergence expression

$$\mathcal{C} = g^r (\mathcal{D}_{sr})^\dagger (\lambda^s) - \lambda^s \mathcal{D}_{sr} g^r, \tag{6.74}$$

which gives one conservation law for every $\mathbf{g}$ allowed by the constraints (6.71). Specifically, the conservation law is

$$\mathcal{C} = g^r \, \mathrm{S}_{-\mathbf{J}}\left(\frac{\partial Q^\alpha(\mathbf{n}, [\mathbf{u}; \mathbf{g}])}{\partial g^r_{\mathbf{J}}} \, \mathbf{E}_{u^\alpha}(L)\right),$$

which is equivalent to

$$\widetilde{\mathcal{C}} = g^r_{\mathbf{J}}\left(\frac{\partial Q^\alpha(\mathbf{n}, [\mathbf{u}; \mathbf{g}])}{\partial g^r_{\mathbf{J}}} \, \mathbf{E}_{u^\alpha}(L)\right). \tag{6.75}$$

Provided that $\mathbf{Q}$ is homogeneous in $[\mathbf{g}]$, the conservation law (6.75) is a multiple of the conservation law (6.63) that is obtained from Noether's (first) Theorem. Hence the results for constrained arbitrary functions bridge the gap between Noether's two theorems on variational symmetries.

Example 6.23 The Euler–Lagrange equations for the Lagrangian

$$L = (u_{1,0} + v_{0,1})^2 + 2(u_{1,0} + v_{0,1})(w_{1,0} + z_{0,1}) - (u - w)(v - z)$$

are

$$\mathbf{E}_u(L) \equiv 2(u + v_{-1,1} + w + z_{-1,1}) - v + z = 0,$$
$$\mathbf{E}_v(L) \equiv 2(u_{1,-1} + v + w_{1,-1} + z) - u + w = 0,$$
$$\mathbf{E}_w(L) \equiv 2(u + v_{-1,1}) + v - z = 0,$$
$$\mathbf{E}_z(L) \equiv 2(u_{1,-1} + v) + u - w = 0.$$

There is a variational symmetry generator,

$$X = \mathsf{g}^1(m, n)(\partial_u + \partial_w) + \mathsf{g}^2(m, n)(\partial_v + \partial_z),$$

for every pair of functions g^1, g^2 that satisfy $\mathsf{g}^1_{1,0} + \mathsf{g}^2_{0,1} = 0$, so the augmented Lagrangian (6.72) is

$$\hat{L} = \mathsf{g}^1\{\mathbf{E}_u(L) + \mathbf{E}_w(L)\} + \mathsf{g}^2\{\mathbf{E}_v(L) + \mathbf{E}_z(L)\} - \lambda(\mathsf{g}^1_{1,0} + \mathsf{g}^2_{0,1}).$$

Taking variations with respect to g^1 and g^2 yields the identities

$$\mathbf{E}_u(L) + \mathbf{E}_w(L) - \lambda_{-1,0} \equiv 0,$$
$$\mathbf{E}_v(L) + \mathbf{E}_z(L) - \lambda_{0,-1} \equiv 0. \tag{6.76}$$

We can eliminate λ from (6.76) to obtain the difference relation

$$S_m\{\mathbf{E}_u(L) + \mathbf{E}_w(L)\} - S_n\{\mathbf{E}_v(L) + \mathbf{E}_z(L)\} \equiv 0;$$

the remainder of (6.76) amounts to

$$\lambda = 4(u_{1,0} + v_{0,1}) + 2(w_{1,0} + z_{0,1}). \tag{6.77}$$

This gives a conservation law for every g^1, g^2 that satisfy the constraint, namely

$$\widetilde{C} = (S_m - I)\left(\mathsf{g}^2_{-1,0}\,\lambda\right) + (S_n - I)\left(\mathsf{g}^1_{0,-1}\,\lambda\right),$$

where λ is given by (6.77). ▲

Notes and further reading

The central theme of this book is that many ideas and techniques for dealing with solutions of differential equations are also applicable, with suitable modification, to difference equations. This is most clearly seen for conservation laws; indeed, the current chapter largely mirrors the presentation in Olver

(1993) of the corresponding results for PDEs. The natural splitting between independent and dependent variables gives a sequence of linear spaces and maps called the *variational complex*. The variational complex for difference equations (see Hydon and Mansfield, 2004) has exactly the same formal structure as its counterpart for differential equations (Olver, 1993). In particular, the cohomology of each complex classifies the equivalence classes of divergences, Lagrangians and Euler–Lagrange expressions. This also leads to the solution of the inverse problem of the calculus of variations, which asks: when is a differential or difference equation an Euler–Lagrange equation?

If the components of a conservation law of a given PDE (in Kovalevskaya form) depend only on a set of initial variables, the characteristic also depends only on the initial variables. Thus, the characteristic is the analogue of the root. This is not the case for PΔEs, for which the root will not generally be a conservation law characteristic. Nevertheless, if a root is known, it can be used to reconstruct a characteristic, from which one can obtain the components of a conservation law. This is a lengthy process that should not be attempted without the aid of computer algebra (see Grant and Hydon, 2013); in practice, it is usually easier to construct the conservation law directly.

Discrete integrable systems have infinite hierarchies of conservation laws that can be obtained in many different ways other than the direct approach (Rasin and Hydon, 2007a). These include a method that exploits Lax pairs (Zhang et al., 2013), the Gardner method (Rasin, 2010), recursion operators (Mikhailov et al., 2011b) and co-recursion operators, which generate a hierarchy of conservation law characteristics (Mikhailov et al., 2011a).

Noether's two theorems on variational symmetries have an interesting history. Brading (2002) and Kosmann-Schwarzbach (2011) are accessible accounts of the mathematics and physics related to these theorems. Surprisingly, Noether's Second Theorem is still not widely known. For a fuller discussion of the difference version of Noether's Theorem, see Dorodnitsyn (2001, 2011). Hydon and Mansfield (2011) introduced the difference analogue of Noether's Second Theorem, together with the intermediate case involving an infinite group subject to differential or difference constraints.

Continuous variational problems may be discretized to yield a *variational integrator*. This approach to finite difference approximation is reviewed in Marsden and West (2001). In effect, a finite difference discretization is a cellular decomposition of the underlying continuous space; this yields an abstract cell complex that amounts to a de Rham complex on lattice space (see Mansfield and Hydon, 2008; Casimiro and Rodrigo, 2012). If a discretization preserves a variational symmetry, the discrete version of Noether's Theorem guarantees that the corresponding conservation law is also preserved.

Exercises

6.1 Find all conservation laws of the form (6.20) for the PΔE

$$u_{1,1} = \frac{u_{0,1}(u_{1,0} - 1)}{u} + 1.$$

Use a first integral to find the general solution of this PΔE.

6.2 Show that the discrete Lotka–Volterra equation (6.34) has five linearly independent conservation laws with components (6.20). [This is a remarkably high number of independent three-point conservation laws for a PΔE of the form (6.19) that is not linearizable.]

6.3 Investigate the symmetries, conservation laws and general solution of

$$u_{1,1} = \frac{1 - uu_{0,1}}{u_{1,0}}.$$

[Look at some low-order generalized symmetries, not just at point symmetries.] How are they related?

6.4 Find all conservation laws (up to equivalence) that are of the form (6.20) for

$$u_{1,1} = -u - \frac{1}{u_{1,0} u_{0,1}}.$$

Mikhailov and Xenitidis (2013) includes a detailed study of this PΔE, its symmetries and conservation laws.

6.5 (a) Show that every solution of the cross-ratio equation (5.74) is a solution of the Toda-type equation (6.56).
    (b) Use (6.64) to find a conservation law of the cross-ratio equation, and determine whether or not it is trivial.
    (c) Use the direct method to find low-order conservation laws of (5.74) and (6.56). To what extent are the conservation laws of these two equations linked?

6.6 Write each of the conservation laws in Example 6.7 as a divergence expression, $C_i(m, n, \mathcal{I}, [\mathcal{A}])$, where $\mathcal{A} = u_{1,1} - u - 1/(u_{1,0} - u_{0,1})$, checking that $C_i(m, n, \mathcal{I}, [0]) \equiv 0$. Hence calculate the root (with respect to $\mathcal{I}$) for each conservation law.

6.7* The autonomous dpKdV equation has the generalized symmetry generator $X = Q_6 \partial_u$, where Q_6 is given in (5.64). Apply this to the conservation law whose components are given in (6.33) and calculate the root of the resulting conservation law. [Note: the calculations are not easy to do by hand, so a computer algebra package should be employed.] Is the 'new' conservation law independent of the conservation laws that are listed in Examples 6.6 and 6.7?

6.8   Show that any conservation law remains a divergence expression under any valid change of lattice coordinates. Why is this divergence expression a conservation law? Prove that the transformed conservation law is trivial if and only if the original conservation law is trivial.

6.9* (a) Calculate the root of the conservation law (F_3, G_3) in Example 6.11, when the dLV equation is written in the form

$$\mathcal{A} \equiv (u_{1,0} - 1)/u - (u_{0,1} - 1)/u_{1,1} = 0.$$

(b) Extend your result to a general invertible transformation of the principal variable: $\tilde{\mathcal{A}} = \theta(\mathbf{n}, \mathcal{I}, \mathcal{A})$, where $\tilde{\mathcal{A}} = 0$ when $\mathcal{A} = 0$ (and both forms of the dLV equation satisfy the maximal rank condition).
(c) Investigate the effect on the root of a valid change of discrete coordinates, bearing in mind that the initial lines will also be transformed.
(d) Generalize your results to all (sufficiently well-behaved) scalar PΔEs that can be written in Kovalevskaya form.

6.10 This problem deals with the condition under which a prolonged vector field X is a variational symmetry generator for a given Lagrangian, L. [The domain is assumed to be $\mathbb{Z}^N$.]

(a) Suppose that L involves only a single dependent variable, u, and let $\mathcal{A} \equiv \mathbf{E}_u(L)$. Prove that if X has the characteristic Q, it is a variational symmetry generator if and only if

$$X(\mathcal{A}) + \mathbf{S}_{-\mathbf{J}}\left(\mathcal{A}\,\frac{\partial Q}{\partial u_{\mathbf{J}}}\right) \equiv 0. \tag{6.78}$$

[Hint: the identities (3.101) and (6.42) are helpful.]
(b) Extend the above result to systems of Euler–Lagrange equations[5], using index notation to state the condition that generalizes (6.78).

6.11 Rearrange the Euler–Lagrange equations (6.58) and hence derive their general solution, (6.59).

---

[5] This question does not address the circumstances under which a given PΔE is a system of Euler–Lagrange equations. (See Notes and further reading in this chapter for details.)

6.12* There exist Lagrangians whose Euler–Lagrange equations are incompatible, in that they have no solutions. A simple example is the innocent-looking Lagrangian

$$L = u_{1,0} \exp(v_{0,1}) + \tfrac{1}{2}\left(\exp(2v) - u^2\right). \qquad (6.79)$$

(a) Show that the Euler–Lagrange equations for (6.79) are incompatible.

(b) Show that if $\exp(v)$ in (6.79) is replaced by w, the Euler–Lagrange equations $\mathbf{E}_u(L) = 0$ and $\mathbf{E}_w(L) = 0$ are compatible. Find all solutions of this system.

(c) Calculate the system of Euler–Lagrange equations for a Lagrangian of the form $L(u, v, u_{1,0}, v_{0,1})$ that is quadratic in each of its arguments. Under what circumstances does this system have no solutions other than $u = v = 0$?

6.13 Write down the Euler–Lagrange equation for the variational problem whose Lagrangian is

$$L = (u_{1,1} + u_{1,0})(u + u_{0,1})(u_{1,0} - u_{0,1}).$$

Show that $X = (-1)^{m+n}\partial_u$ generates variational symmetries and apply Noether's Theorem to obtain the corresponding conservation law.

6.14 (a) Write down the Euler–Lagrange equations for the Lagrangian

$$L = (v_{1,1} - u)(v_{0,1} - u_{1,0}) + \ln|u - v|.$$

(b) Show that every Lie point symmetry of the Euler–Lagrange equations is a variational symmetry. [The Lie algebra of point symmetry generators is five-dimensional.]

(c) Use Noether's Theorem to determine the conservation laws that correspond to the above variational symmetries.

6.15 This question examines the variational problem on $\mathbb{Z}^2$ whose Lagrangian is

$$L = \frac{u_{1,0}}{u v_{0,1}} + \ln|v|.$$

(a) Write down the Euler–Lagrange equations and show that both of these are satisfied if and only if

$$v_{0,1} = \frac{u_{1,0}}{u}. \qquad (6.80)$$

(b) Find all Lie point symmetry generators for (6.80) and determine which of these generate variational symmetries.

(c) Use the proof of Noether's Second Theorem to calculate a difference relation that links the Euler–Lagrange equations and check that this relation holds.

6.16* [Note: Before attempting this exercise, you must be able to use a metric to raise and lower tensor indices. The exercise is written in standard notation for electromagnetism.]

The interaction of a scalar particle of mass m and charge e with an electromagnetic field has a variational formulation with gauge symmetries that depend on one arbitrary function. The independent variables are the standard flat space-time coordinates $\mathbf{x} = (x^0, x^1, x^2, x^3)$, where x^0 denotes time. The dependent variables are the complex-valued scalar ψ (which is the wavefunction for the particle), its complex conjugate, ψ^*, and the real-valued electromagnetic four-potential A^μ, $\mu = 0, 1, 2, 3$; indices are raised and lowered using the metric $\eta = \mathrm{diag}\{-1, 1, 1, 1\}$. For details of the continuous problem, see Christiansen and Halvorsen (2011), which introduces the following finite difference approximation that preserves the gauge symmetries. The mesh $\mathbf{x}(\mathbf{n})$ is uniformly spaced in each direction, with step lengths

$$h^\mu = (S_\mu - I)(x^\mu(\mathbf{n})), \qquad \mu = 0, \dots, 3.$$

It is useful to introduce the scaled forward difference operators

$$\overline{D}_\mu = \frac{S_\mu - I}{h^\mu}, \qquad \mu = 0, \dots, 3,$$

so that $\overline{D}_\mu$ tends to the total derivative with respect to x^μ in the limit as h^μ tends to zero; note that

$$\overline{D}_\mu^\dagger = -\frac{I - S_\mu^{-1}}{h^\mu}.$$

Let $\psi(\mathbf{n})$ denote the approximation to the wavefunction ψ at $\mathbf{x}(\mathbf{n})$, and let $A_\mu(\mathbf{n})$ denote the approximation to the (lowered) four-potential A_μ on the edge that connects the points $\mathbf{x}(\mathbf{n})$ and $S_\mu \mathbf{x}(\mathbf{n})$. Up to a sign, the discretized Lagrangian is

$$L = \tfrac{1}{4} F_{\rho\mu} F^{\rho\mu} + \left(\nabla_\mu \psi\right)(\nabla^\mu \psi)^* + m^2 \psi\psi^*,$$

where, for each $\rho, \mu = 0, \dots, 3$,

$$F_{\rho\mu} = \overline{D}_\rho A_\mu - \overline{D}_\mu A_\rho.$$

and

$$\nabla_\mu = \frac{1}{h^\mu} \{ S_\mu - \exp(-ieh^\mu A_\mu) I \} = \overline{D}_\mu + \frac{1}{h^\mu} \{ 1 - \exp(-ieh^\mu A_\mu) \} I.$$

(a) Write down the Euler–Lagrange equations corresponding to variations in ψ, ψ^* and A^σ.

(b) Show that the variational symmetries include the gauge symmetries generated by

$$X = -ie\psi g \partial_\psi + ie\psi^* g \partial_{\psi^*} + \eta^{\alpha\tau} \left(\overline{D}_\alpha g \right) \partial_{A^\tau},$$

where $g(\mathbf{n})$ is arbitrary. (It turns out that these are the only gauge symmetries.)

(c) Use Noether's Second Theorem to determine the relationship between the Euler–Lagrange equations, and verify that this is borne out by your answer to (a).

(d) Investigate what happens to the above results in the limit as h^0 tends to zero (giving a discretization in space only) and in the limit as all h^μ tend independently to zero.

6.17 (a) Find all Lie point symmetry generators of (6.58) and identify which of these generate variational symmetries.

(b) Use the method of §6.8 to determine a set of conservation laws that depend on an arbitrary function that is subject to a single constraint.

6.18 The *lattice KdV equation* is

$$u_{1,1} - u - \frac{1}{u_{1,0}} + \frac{1}{u_{0,1}} = 0. \tag{6.81}$$

This is not an Euler–Lagrange equation, but it can be turned into one by introducing a potential, v, such that

$$u = v_{0,-1} - v_{-1,0}.$$

(a) Find a Lagrangian, $L(m, n, [v])$, whose Euler–Lagrange equation is equivalent to (6.81).

(b) By definition, u is unaffected by transformations of the form

$$v \mapsto v + \varepsilon g(m, n) \quad \text{such that} \quad g_{0,-1} = g_{-1,0},$$

so these are symmetries of the Euler–Lagrange equation. Are they variational symmetries? If not, seek a new Lagrangian for which these are variational symmetries.

(c) Construct the augmented Lagrangian and use it to obtain conservation laws of the Euler–Lagrange equation.

(d) What is the connection between the lattice KdV equation and the dpKdV equation?

6.19 Investigate the symmetries, conservation laws and general solution of the Euler–Lagrange equations corresponding to the Lagrangian

$$L = \frac{u_{1,0} + v_{0,1}}{u + v} \, .$$

References

Abramowitz, M. and Stegun, I. A. 1965. *Handbook of mathematical functions.* New York: Dover.

Adler, V. E., Bobenko, A. I., and Suris, Y. B. 2003. Classification of integrable equations on quad-graphs. The consistency approach. *Comm. Math. Phys.*, **233**, 513–543.

Adler, V. E., Bobenko, A. I., and Suris, Y. B. 2009. Discrete nonlinear hyperbolic equations: classification of integrable cases. *Funktsional. Anal. i Prilozhen.*, **43**, 3–21.

Adler, V. E., Bobenko, A. I., and Suris, Y. B. 2012. Classification of integrable discrete equations of octahedron type. *Int. Math. Res. Not.*, 1822–1889.

Bluman, G. W. and Anco, S. C. 2002. *Symmetry and integration methods for differential equations.* New York: Springer.

Bobenko, A. I. and Suris, Y. B. 2008. *Discrete differential geometry: integrable structure.* Providence, RI: American Mathematical Society.

Brading, K. A. 2002. Which symmetry? Noether, Weyl, and conservation of electric charge. *Stud. Hist. Philos. Sci. B Stud. Hist. Philos. Modern Phys.*, **33**, 3–22.

Bridgman, T., Hereman, W., Quispel, G. R. W., and van der Kamp, P. H. 2013. Symbolic computation of Lax pairs of partial difference equations using consistency around the cube. *Found. Comput. Math.*, **13**, 517–544.

Budd, C. J. and Piggott, M. D. 2003. Geometric integration and its applications. In *Handbook of numerical analysis*, Vol. XI, pp. 35–139. Amsterdam: North-Holland.

Casimiro, A. C. and Rodrigo, C. 2012. First variation formula and conservation laws in several independent discrete variables. *J. Geom. Phys.*, **62**, 61–86.

Christiansen, S. H. and Halvorsen, T. G. 2011. Discretizing the Maxwell–Klein–Gordon equation by the lattice gauge theory formalism. *IMA J. Numer. Anal.*, **31**, 1–24.

Dorodnitsyn, V. 2001. Noether-type theorems for difference equations. *Appl. Numer. Math.*, **39**, 307–321.

Dorodnitsyn, V. 2011. *Applications of Lie groups to difference equations.* Boca Raton, FL: CRC Press.

Duistermaat, J. J. 2010. *Discrete integrable systems: QRT maps and elliptic surfaces.* New York: Springer.

Elaydi, S. 2005. *An introduction to difference equations*, 3rd edn. New York: Springer.

Erdmann, K. and Wildon, M. J. 2006. *Introduction to Lie algebras*. Springer Undergraduate Mathematics Series. London: Springer.

Fuchs, J. and Schweigert, C. 1997. *Symmetries, Lie algebras and representations*. Cambridge: Cambridge University Press.

Gaeta, G. 1993. Lie-point symmetries of discrete versus continuous dynamical systems. *Phys. Lett. A*, **178**, 376–384.

Gao, M., Kato, Y., and Ito, M. 2004. Some invariants for kth-order Lyness equation. *Appl. Math. Lett.*, **17**, 1183–1189.

Garifullin, R. N. and Yamilov, R. I. 2012. Generalized symmetry classification of discrete equations of a class depending on twelve parameters. *J. Phys. A*, **45**, 345205, 23pp.

Grammaticos, B., Kosmann-Schwarzbach, Y., and Tamizhmani, T. (eds). 2004. *Discrete integrable systems*. Berlin: Springer.

Grammaticos, B., Halburd, R. G., Ramani, A., and Viallet, C.-M. 2009. How to detect the integrability of discrete systems. *J. Phys. A*, **42**, 454002, 30pp.

Grant, T. J. and Hydon, P. E. 2013. Characteristics of conservation laws for difference equations. *Found. Comput. Math.*, **13**, 667–692.

Gregor, J. 1998. The Cauchy problem for partial difference equations. *Acta Appl. Math.*, **53**, 247–263.

Hairer, E., Lubich, C., and Wanner, G. 2006. *Geometric numerical integration: structure-preserving algorithms for ordinary differential equations*, 2nd edn. Berlin: Springer.

Hardy, G. H. and Wright, E. M. 1979. *An introduction to the theory of numbers*, 5th edn. Oxford: Clarendon Press.

Hirota, R. 1977. Nonlinear partial difference equations. I. A difference analogue of the Korteweg–de Vries equation. *J. Phys. Soc. Japan*, **43**, 1424–1433.

Hydon, P. E. 2000a. Symmetries and first integrals of ordinary difference equations. *Proc. R. Soc. Lond. Ser. A Math. Phys. Eng. Sci.*, **456**, 2835–2855.

Hydon, P. E. 2000b. *Symmetry methods for differential equations: a beginner's guide*. Cambridge: Cambridge University Press.

Hydon, P. E. and Mansfield, E. L. 2004. A variational complex for difference equations. *Found. Comput. Math.*, **4**, 187–217.

Hydon, P. E. and Mansfield, E. L. 2011. Extensions of Noether's second theorem: from continuous to discrete systems. *Proc. R. Soc. Lond. Ser. A Math. Phys. Eng. Sci.*, **467**, 3206–3221.

Iserles, A., Munthe-Kaas, H. Z., Nørsett, S. P., and Zanna, A. 2000. Lie-group methods. *Acta Numer.*, **9**, 215–365.

Kelley, W. G. and Peterson, A. C. 2001. *Difference equations: an introduction with applications*, 2nd edn. San Diego, CA: Harcourt/Academic Press.

Kim, P. and Olver, P. J. 2004. Geometric integration via multi-space. *Regul. Chaotic Dyn.*, **9**, 213–226.

Kosmann–Schwarzbach, Y. 2011. *The Noether theorems: Invariance and conservation laws in the twentieth century*, Translated, revised and augmented from the 2006 French edition by B. E. Schwarzbach. Sources and Studies in the History of Mathematics and Physical Sciences. New York: Springer.

Leimkuhler, B. and Reich, S. 2004. *Simulating Hamiltonian dynamics*. Cambridge: Cambridge University Press.

Levi, D. and Yamilov, R. I. 2011. Generalized symmetry integrability test for discrete equations on the square lattice. *J. Phys. A*, **44**, 145207, 22pp.

Levi, D., Vinet, L., and Winternitz, P. 1997. Lie group formalism for difference equations. *J. Phys. A*, **30**, 633–649.

Maeda, S. 1987. The similarity method for difference equations. *IMA J. Appl. Math.*, **38**, 129–134.

Mansfield, E. L. and Hydon, P. E. 2008. Difference forms. *Found. Comput. Math.*, **8**, 427–467.

Mansfield, E. L. and Szanto, A. 2003. Elimination theory for differential difference polynomials. In *Proceedings of the 2003 International Symposium on Symbolic and Algebraic Computation*, pp. 191–198. New York: ACM.

Marsden, J. E. and West, M. 2001. Discrete mechanics and variational integrators. *Acta Numer.*, **10**, 357–514.

McLachlan, R. and Quispel, R. 2001. Six lectures on the geometric integration of ODEs. In *Foundations of computational mathematics (Oxford, 1999)*, pp. 155–210. Cambridge: Cambridge University Press.

Mickens, R. E. 1990. *Difference equations: theory and applications*, 2nd edn. New York: Van Nostrand Reinhold.

Mikhailov, A. V. and Xenitidis, P. 2013. Second order integrability conditions for difference equations: an integrable equation. *Lett. Math. Phys.*, DOI: 10.1007/s11005-013-0668-8.

Mikhailov, A. V., Wang, J. P., and Xenitidis, P. 2011a. Cosymmetries and Nijenhuis recursion operators for difference equations. *Nonlinearity*, **24**, 2079–2097.

Mikhailov, A. V., Wang, J. P., and Xenitidis, P. 2011b. Recursion operators, conservation laws, and integrability conditions for difference equations. *Theor. Math. Phys.*, **167**, 421–443.

Olver, P. J. 1993. *Applications of Lie groups to differential equations*, 2nd edn. New York: Springer.

Olver, P. J. 1995. *Equivalence, invariants, and symmetry*. Cambridge: Cambridge University Press.

Ovsiannikov, L. V. 1982. *Group analysis of differential equations*. Translated from the Russian by Y. Chapovsky, Translation edited by W. F. Ames. New York: Academic Press.

Papageorgiou, V. G., Nijhoff, F. W., and Capel, H. W. 1990. Integrable mappings and nonlinear integrable lattice equations. *Phys. Lett. A*, **147**, 106–114.

Quispel, G. R. W. and Sahadevan, R. 1993. Lie symmetries and the integration of difference equations. *Phys. Lett. A*, **184**, 64–70.

Quispel, G. R. W., Nijhoff, F. W., Capel, H. W., and van der Linden, J. 1984. Linear integral equations and nonlinear difference-difference equations. *Phys. A*, **125**, 344–380.

Quispel, G. R. W., Roberts, J. A. G., and Thompson, C. J. 1988. Integrable mappings and soliton equations. *Phys. Lett. A*, **126**, 419–421.

Rasin, A. G. 2010. Infinitely many symmetries and conservation laws for quad-graph equations via the Gardner method. *J. Phys. A*, **43**, 235201, 11pp.

Rasin, A. G. and Schiff, J. 2013. The Gardner method for symmetries. *J. Phys. A*, **46**, 155202, 15pp.

Rasin, O. G. and Hydon, P. E. 2007a. Conservation laws for integrable difference equations. *J. Phys. A*, **40**, 12763–12773.

Rasin, O. G. and Hydon, P. E. 2007b. Symmetries of integrable difference equations on the quad-graph. *Stud. Appl. Math.*, **119**, 253–269.

Stephani, H. 1989. *Differential equations: their solution using symmetries*. Cambridge: Cambridge University Press.

van der Kamp, P. H. 2009. Initial value problems for lattice equations. *J. Phys. A*, **42**, 404019, 16pp.

Wilf, H. S. 2006. *generatingfunctionology*, 3rd edn. Wellesley, MA: A. K. Peters. The second edition may be downloaded (subject to conditions) from the author's website.

Xenitidis, P. and Nijhoff, F. 2012. Symmetries and conservation laws of lattice Boussinesq equations. *Phys. Lett. A*, **376**, 2394–2401.

Zhang, D.-J., Cheng, J.-W., and Sun, Y.-Y. 2013. Deriving conservation laws for ABS lattice equations from Lax pairs. *J. Phys. A*, **46**, 265202, 19pp.

Index